Praise for

Essential Cordwood Building

Rob Roy is the world's foremost expert on cordwood construction and his new book is the best and most comprehensive guide to the subject ever produced.

— Lloyd Kahn, author of *Shelter, Tiny Homes, and Small Homes*

Rob Roy has influenced a whole regeneration with his cordwood building techniques, liberated countless people from the slavery of a large mortgage, and enabled ordinary people to build all kinds of structures from workshops to comfortable homes in low impact, ecological ways. Here he has gathered the latest methods with the essence of these cordwood techniques. Full of information and stories. A great read.

— Maddy Harland, Editor, Co-Founder of *Permaculture Magazine*,
Author of *Fertile Edges — regenerating land, culture and hope*

Cordwood building has always been my favorite, and it is obviously Rob Roy's too. Here is a mine of information, laid out well, covering all aspects of cordwood building. He looks at the overall considerations, then gets down in the detail, always practical, and writes simply, with lightness and humor, from his long experience. There is more than enough here to get you started down a long, creative, satisfying path as a self-builder.

— Tony Wrench lives with his partner Faith in a cobwood roundhouse
he designed and they built together 20 years ago in West Wales, UK.
He runs occasional natural building courses and is the
author of *Building a Low Impact Roundhouse*

Rob Roy's *Essential Cordwood Building* certainly contains all of the information that a newcomer would need to delve confidently into a wide variety of cordwood projects. Rob has been the go-to expert on cordwood construction at www.greenhomebuilding.com for well over a decade and has always answered reader's questions with authority, wit and patience. I could trust nobody more implicitly than Rob to teach the attributes of what has to be one of the most effective, economical, ecological, energy efficient and beautiful approaches to natural building. In one simple operation you can build walls that provide both insulation and thermal mass, requiring no further interior or exterior finish work; there is no other method where this is true!

— Kelly Hart, founder of www.greenhomebuilding.com

In *Essential Cordwood Building,* Rob Roy has distilled the critical information required to think and plan your way through the design and construction of a cordwood structure, and presented it in a very accessible format. His tone is familiar and personal, and draws upon a deep wealth of experience in his craft — there may well be no better resource on cordwood construction than Rob Roy! The book is well-organized and thorough in its detail, proving to be a truly valuable resource to anyone considering working with cordwood.

— Jacob Deva Racusin, Co-Owner, New Frameworks Natural Design/Build

With *Essential Cordwood Building,* Rob Roy has succeeded, as only he can, in combining an articulate step by step methodology with an enduring lesson in personal economics. If you want the single biggest investment of your life to be a get-it-done-with-and-move-on experience — free of those mortgage entanglements and full of log-and-mortar security — you'd do well to follow the path that Rob Roy's has so clearly marked, and build your own cordwood home.

— Richard Freudenberger, Energy and Resource Manager at Living Web Farms and former publisher, *BackHome Magazine*

essential
CORDWOOD BUILDING

sustainable
building
essentials

essential CORDWOOD BUILDING

the complete **step-by-step** guide

Rob Roy

new society
PUBLISHERS
www.newsociety.com

New Society
Sustainable Building Essentials Series

Series editors
Chris Magwood and Jen Feigin

Title list
Essential Hempcrete Construction, Chris Magwood
Essential Prefab Straw Bale Construction, Chris Magwood
Essential Building Science, Jacob Deva Racusin
Essential Light Straw Clay Construction, Lydia Doleman
Essential Sustainable Home Design, Chris Magwood
Essential Cordwood Building, Rob Roy

See www.newsociety.com/SBES for a complete list of new and forthcoming series titles.

THE SUSTAINABLE BUILDING ESSENTIALS SERIES covers the full range of natural and green building techniques with a focus on sustainable materials and methods and code compliance. Firmly rooted in sound building science and drawing on decades of experience, these large-format, highly illustrated manuals deliver comprehensive, practical guidance from leading experts using a well-organized step-by-step approach. Whether your interest is foundations, walls, insulation, mechanical systems, or final finishes, these unique books present the essential information on each topic including:

- Material specifications, testing, and building code references
- Plan drawings for all common applications
- Tool lists and complete installation instructions
- Finishing, maintenance, and renovation techniques
- Budgeting and labor estimates
- Additional resources

Written by the world's leading sustainable builders, designers, and engineers, these succinct, user-friendly handbooks are indispensable tools for any project where accurate and reliable information is key to success. GET THE ESSENTIALS!

Cover design by Diane McIntosh.
Illustrations by Dale Brownson.
Background photo AdobeStock_58151515.
All interior photos by Rob or Jaki Roy unless otherwise noted.

Printed in Canada. First printing November 2017.

Any other inquiries can be directed by mail to:
New Society Publishers
P.O. Box 189, Gabriola Island, BC V0R 1X0, Canada
(250) 247-9737

Library and Archives Canada Cataloguing in Publication

Roy, Robert L., author
Essential cordwood building : the complete step-by-step guide / Rob Roy.

(Sustainable building essentials)
Includes index.
Issued in print and electronic formats.
ISBN 978-0-86571-852-4 (softcover).--ISBN 978-1-55092-647-7 (PDF).--
ISBN 978-1-77142-242-0 (EPUB)

1. Log-end houses--Design and construction--Handbooks, manuals, etc. 2. Building, Wooden--Handbooks, manuals, etc. 3. Sustainable construction--Handbooks, manuals, etc. I. Title. II. Series: Sustainable building essentials

TH4818.W6R68 2018 694 C2017-905404-X
C2017-905405-8

Funded by the Government of Canada
Financé par le gouvernement du Canada
Canada

New Society Publishers' mission is to publish books that contribute in fundamental ways to building an ecologically sustainable and just society, and to do so with the least possible impact on the environment, in a manner that models this vision.

FSC www.fsc.org MIX Paper from responsible sources FSC® C016245
Certified B Corporation
new society PUBLISHERS www.newsociety.com

Contents

Acknowledgments

THIS BOOK WOULD NOT HAVE BEEN POSSIBLE without the work of many cordwood builders around the world. In particular, I was greatly assisted by Bruce Kilgore and Nancy Dow, Alan Stankevitz, Jim Juczak, Tom Huber, Cliff Shockey, Richard Flatau, Sandy Clidaras, Dr. Kris Dick, P.E., and Ianto Evans and Linda Smiley for help with cobwood, and Peter Robey and Blythe Tait in Tasmania for sharing their experience with "rendering" (plastering) cordwood.

Thanks, also, to Rob West at New Society for asking me to do this new book and helping to guide me through the format, my editor Linda Glass, graphic designer Mary Jane Jessen, friends Judith Plant, Ingrid Witvoet , Sue Custance, and all the NSP gang who played a part in bringing this to print.

Chapter 1

Introduction

What Is Cordwood Masonry?

Cordwood masonry (sometimes called *stackwall* construction in Canada) is a term describing the construction of exterior or interior walls out of short logs — *log-ends* — laid transversely in the wall and supported by an insulated mortar matrix. The mortar portion of the wall can be made with cement- or lime-based mortars, cob (clay, straw, and sand), "papercrete," or — a new development — *hempcrete.* The walls can be load-bearing in non-seismic zones, but are more commonly used as infilling within a strong timber frame. A relatively small number of cordwood homes — probably less than five percent — use a *double-wall* technique: separate interior and exterior cordwood walls with the space between them completely filled with insulation, as described in Chapter 6.

Cordwood masonry has a long history, which I discussed in my book *Cordwood Building: A Comprehensive Guide to the State of the Art* (New Society Publishers, 2016). There are existing cordwood buildings in North America and Europe dating back to the 19th century, which are documented in that book. More recently, cordwood has spread to South and Central America, and has enjoyed a rebirth in Scandinavia as well as in Britain, where cob has become a popular alternative for mortar amongst the natural builders.

I have recently learned of an exciting new development in cordwood's history, a site called Slawenburg Raddush (Slavic Fort Raddush), near the town of Vetschau in the German federal state of Brandenburg. Raddush was originally built around the 9th or 10th century AD, employing log and cordwood techniques on the inner and outer surfaces of the massive walls. It was still clearly recognizable as a ring-shaped wooden structure in the early 20th century. The fort was reconstructed during the 1990s using the

1.1: *Raddush Fort in eastern Germany.*

1.2: *Raddush Fort, cordwood detail.*

1.3: *Raddush Fort, interior wall.*

original techniques. See Figures 1.1, 1.2, and 1.3. The internal cavity between the wooden walls was filled with sand, earth, and clay, whereas today we use some form of insulation. Raddush can be visited today and houses a museum, a conference room, and a restaurant. The website is www.slawenburg-raddusch.de/english/.

Chapter 2

Rationale

Why Build with Cordwood Masonry?

SINCE THE 1980s, I have been answering this fundamental question with my "5-E" list of cordwood masonry advantages. It still holds true:

1. *Economy.* Cordwood masonry walls are low in cost, particularly when the owner-builder has a local source of appropriate wood. If clay is readily available on site, *cobwood* construction is an option, saving on Portland and lime. Sand and sawdust (used as insulation and/or as a cement retarder) can usually be bought quite inexpensively. Sand might even be indigenous to the building site.
2. *Energy Efficiency.* Built properly, and with a wall thickness appropriate to the local climate and building size, cordwood homes are easy to heat in the winter and keep cool during the summer.
3. *Easy to Build.* Children, grandmothers, and beavers can all build with cordwood masonry ... and have done so time and again. Our oldest son, Rohan, built his first little cordwood playhouse at age seven and was teaching cordwood masonry to Chicago's inner city youth when he was nine. His brother Darin grew up with cordwood, has taught it with us at Earthwood, and built Driftwood, his own cordwood home.
4. *Esthetically Pleasing.* "A cordwood wall combines the warmth of wood with the pleasing relief and visual interest of stone masonry." I wrote those words in 1992. It's still true, but build quality is getting better all the time. Many builders have taken cordwood to an art form in the past ten years or so.
5. *Environmentally Friendly.* Cordwood makes use of wood which might otherwise go to waste — even tipped into landfills. I have used ends and pieces from sawmills, log cabin manufacturers, and furniture makers. A hollow log is not much use at the sawmill, but it can be an interesting feature in a cordwood wall.

Mortgage? Or Mortgage Freedom?

Cordwood buildings have been built for next to nothing (our Hermit's Hut guesthouse cost less than $1,000 in 2011) to many millions (the architect-designed and contractor-built 10,000-square foot Arcus Center at Kalamazoo College came in at around $5,000,000). Most cordwood homes have been built without a mortgage, including some big, beautiful energy-efficient ones like Bruce and Nancy's Ravenwood, Alan Stankevitz's two-story hexadecagon in Minnesota, as well as our own Earthwood home, and Mushwood, our lake cottage. In a recent phone conversation, Alan and I compared notes on what we knew of cordwood builders' home ownership situations. Offhand, we could not think of any with a mortgage. But why? Well, banks may be reluctant to loan money to owner-builders, especially for a building style out of the mainstream. Or, maybe cordwood builders don't need a mortgage. They own a piece of land and adopt a pay-as-they-go (or proceed-as-they-can-afford) strategy. Alan commented that cordwood makes "a superior house, costs less than most others, and is environmentally sound."

Common strategies that enable debt-free cordwood home ownership (and this is probably true with other natural building methodologies as well) are: (1) already owning the land, (2) building it yourself, (3) making use of indigenous and recycled materials, (4) paying for materials as you go, (5) keeping it small, and (6) keeping it simple. Watch out for 5 and 6, though: A small house can be hopelessly complicated — therefore expensive — whereas a large house could be of simple design, saving time and money. With regard to (3), Alan said "Once people found out I was doing cordwood, they really got into it. They wanted to help. And they'd tell me where I could get good materials for little money."

Widely Varying Costs

I hesitate to give actual cost estimates for a cordwood home. The one truth I have learned in 70 years is this: *Everybody's different.* Their abilities to save and budget are different. Their talent for "cultivating coincidences" (procuring materials) is different. Their design aspirations — for size and complexity — are different. Their access to indigenous materials is different.

The reality is that the cordwood masonry is not a very large part of the home's material cost. Foundation, roofing, stairs, heating, electric, and plumbing systems: these are all — or can be — big ticket items common to any style of home.

Two areas where a lot of money is spent, even on owner-built housing, are kitchens (cabinets, countertop, sinks, etc.) and doors and windows. But Jaki and I have always saved a tremendous amount by making our own doors, getting perfectly good thermal pane windows from the "back room" of local manufacturers, and watching for deals when people "upgrade" their kitchens — tearing out perfectly good cabinets, sinks, and countertops in the process.

Where you build makes a huge difference, too. I know of cordwood buildings in Central America — particularly Belize, Guatemala, and Nicaragua — that were built at incredibly low cost and others, in places like Massachusetts and northern California, that ran up a pretty high tab (but these people were not short of money).

Owning the land makes a huge difference. You have to have the land before you can build the house. And it enables you to start with a small practice building, maybe a "temporary shelter" that you can live in while you build the main home, saving shelter costs immediately. You can employ an add-on strategy and pay for expansion as you go. (One caution here, though: Plan for any desired add-ons at the initial design stage. Some buildings are difficult to add on to — round ones, for example.)

I learned the best part of my economic strategies — which has kept us debt-free for 44 years — from two sources: my father and Henry David Thoreau. My father said, "If a man earns $100,000 a year, but spends $110,000, he's poor. If he earns $10,000 and spends $9,000, he's rich." I never forgot that. And from Henry I learned about real — *empiric* — economy. In the first chapter of *Walden,* "Economy," we learn that the "necessaries of life" are food, fuel, shelter, and clothing. These are the things that keep our body temperatures at 98.6 degrees F, and, therefore, healthy. Of these "necessaries," shelter is the single biggest cost — up to 50 percent of expenditure in places like California. Cordwood is cheap. Like me.

I wrote a book called *Mortgage Free!,* which is full of strategies for avoiding mortgage — a word, incidentally, that derives from the Old French meaning, literally, *death pledge.* The book is out of print, but you can find used copies through Amazon, or get an ebook from Chelsea Green Publishing. I will not attempt to rehash its 300 pages here, but will give you my First Law of Empiric Economics:

A dollar saved is worth a whole lot more than a dollar earned, because we have to earn so darned many of them to save so precious few.

Shockingly, a high percentage of Americans are net *negative* savers: they are going into debt. It is rare to find someone who saves (puts away) ten percent of their income. For these people, if they find a dollar on the sidewalk — or save a buck through thrift — what is that dollar actually worth to them? Well, even for the good savers it's worth $10 that they didn't have to earn in order to save it.

In short, cordwood masonry is labor intensive but materials cheap. When you build a cordwood wall, you are simultaneously attending to many things: structure, interior finish, exterior finish, insulation, and a thermal mass not commonly found in conventional walls. With insulated stick-frame, you've got to frame it, sheathe the outside, and apply some sort of siding. On the interior, you have insulation, vapor barrier, sheetrock, taping and spackling, painting three coats ... you could easily have ten different operations to complete the wall. Lay a log-end and it's done.

Ballpark Cost Estimates

The cost of the cordwood itself varies tremendously around North America, and around the planet. It is a building technique that should be employed where wood is plentiful. But, even in those areas, wood procurement costs can vary from "next to nothing" to "quite a bit." Those who have appropriate wood on their own property are very well off in this regard, but there are often other sources of free or nearly free wood. I have gone to sawmills and found wood unsuitable for cutting into lumber — too short, too small in diameter, hollow core — and carried it away in my pickup truck for next to nothing. One cedar log home builder nearby lets me "clean up the yard" once a year; I carry away all the ends and pieces too small for his purposes. It is a great advantage that short pieces — 8 to 24 inches — can be used for cordwood masonry. I have also visited wooden furniture makers for the same reason. Another source is standing deadwood, trees which have died from fire, insects, or encroaching water. After checking that the wood is still sound (i.e. no bugs or rot), strike a deal with the owner.

If you have to purchase wood, you might get it from the same loggers who supply your local sawmills. Or firewood suppliers. In our area, people want hardwood for firewood, and the suppliers know this all too well. But let them know that you are in the market for pine, spruce, fir, or quaking aspen — frowned on in the Northeast as firewood — and they may be happy to supply you with a load.

You will work in *face cords,* which are described in Chapter 5. This is a unit well-known to loggers. The cost of a purchased face cord can vary greatly in different areas, with $60 to $100 dollars per face cord being the current parameters in northern New York.

Sawdust for insulation can also vary quite a bit in price from place to place. In some areas, it has a value as bedding at farms (as it does where I live), but my sawyer still lets me fill up my small pickup truck — I use a snow shovel — for $5 a load. At that rate, insulation costs for a cordwood wall are very low indeed.

The easiest cost estimate component for a cordwood wall is probably the mortar, but, even here, the masonry sand can vary from free — if it is on site, as is the case for quite a few towns near us — to quite a bit, if it has to be delivered some distance. The sand itself is not the big cost here (just $20 a cubic yard in New York); it is usually the transport. For both Portland cement and lime, $11 a bag is a fairly average cost for North America, although it might vary up to 20 percent on either side of that number.

So, with all these variables in mind, we can make a "ballpark" chart for the cost per square foot of completed wall (Table 2.1). Remember, this includes interior finish, exterior finish, structure, and insulation. To create this chart, I used $11 per bag as the cost for both Portland and lime, $40 per cubic yard for masonry sand (delivered), and varying costs for face cords depending on the length of the log-ends (width of the wall): $40 for 8″, $60 for 12″, $80 for 16″ and $120 for 24″. Sawdust is tough to figure because the cost range is so wide, but, for our purposes here, I have used a value of $10 per cubic yard. Mortar and sawdust insulation widths are as per Figure 12.7 in Chapter 12.

These are the approximate materials costs per square foot of cordwood masonry at various wall widths, based on the materials' prices mentioned above. To use the chart, tally the total square footage of all the cordwood masonry in your design, and multiply by the total cost per square foot (the last column). For example, if your little sauna with 8-inch walls has 280 square feet of cordwood, multiply 280 times $2.53 for a materials cost of $708. A small house with 16-inch walls and 800 square feet of cordwood will cost 800 times $4.46, for a total of $3,568. If you have your own wood, the cost of the cordwood masonry can be cut almost in half. You might have your own sand. If you are using cob instead of mortar, and the materials are indigenous to the site, the lime and Portland costs are eliminated. So, the chart reflects a kind of a worst-case scenario. Use your own local dollar figures for greater accuracy, adjusting proportionally.

Table 2.1: Ballpark cost estimates

Wall Width	Lime & Portland	Masonry Sand	Wood Cost	Sawdust Insulation	Total Cost per sq. ft.
8″	$1.14	$0.35	$1.00	$0.04	$2.53
12″	1.38	0.40	1.50	0.04	3.32
16″	1.83	0.57	2.00	0.06	4.46
24″	2.75	0.82	3.00	0.08	6.65

Chapter 3

Appropriate Use

CORDWOOD MASONRY is appropriate for large and small homes and outbuildings, including guesthouses, garages, garden and storage sheds, and playhouses. Cordwood masonry makes an excellent sauna — as hundreds of builders have found. It has also been used to make specialty internal walls in homes, bars, and restaurants.

Except for Antarctica and Africa — as far as I know — cordwood masonry has been used on all of the world's continents and in a variety of climates, wet and dry. Special considerations — to be discussed — must be adhered to in wet climates, but a successful and beautiful cordwood home was built in Mountain View on Big Island, Hawaii, with its 200 inches of rain per year. The home is featured in *Cordwood Building: A Comprehensive Guide to the State of the Art.*

In all cases, care must be taken in the selection and preparation of the wood. Special considerations pertain to the various species of woods, usually a function of the wood's density, and these are discussed in Chapter 5.

Inappropriate Use

Inappropriate use would be a situation where favorable wood must be hauled a long distance to the building site. Therefore, sites in or near forested areas are more appropriate than deserts or prairies. Also, be advised that cordwood may not be the best choice for below-grade applications. We tried that in 1981, but standing rainwater on the unprotected concrete floor caused log-end swelling in the first course of cordwood. As the building (Earthwood) was round, this caused a tilting out of the curved wall as it hinged on the outer mortar joint. Although our problem was exacerbated by the deliberate use of dense hardwood log-ends for thermal mass, backfilling cordwood masonry is rife with potential waterproofing and structural problems.

Many people have built freestanding cordwood masonry walls outside — in parks and on private property — and that seems to have worked. Keep the cordwood off the ground on a concrete or stone foundation, and put some kind of cap with overhang to both sides to prevent constant water draining down the wall. Not taking these two precautions would, indeed, be "inappropriate."

Fire Resistance of Cordwood Masonry

A mostly wooden wall suggests a fire danger. In point of fact, cordwood masonry is very difficult to set on fire. Fire needs fuel, heat, and air. The fuel is there, right enough, but 40 percent of the wall, typically, is massive noncombustible mortar, which takes the heat out of any fire that starts. Once the outer quarter-inch or so of the cordwood has charred, oxygen is greatly cut off, and rapid combustion ceases. "Cordwood Jack" Henstridge had an interior fire in one of the rooms of his home in New Brunswick. The cordwood wall was charred, but the fire went out. No permanent structural damage. I imagine he either sanded the log-ends or plastered over them, I'm not sure.

In his paper "Fire Endurance Test of a Non-Load-Bearing (Cordwood) Wall," appearing in *Cordwood and the Code* (2005), Vincent Hartung reports on testing a cordwood wall for fire endurance at the University of New Brunswick in 1995. It is an illustrated 2,000-word paper and has lots of details about how the test was performed (according to standard testing procedure) that should satisfy any building code enforcement officer on cordwood's ability to resist fire spread. Incredibly, the researchers put "a single propane-powered burner of the hand-held type used to melt roofing materials" against the wall. The highest flame temperature reached was 2,100 degrees Fahrenheit (1,150 degrees Celsius) at 12 inches from the wall. Vince concludes his paper with this: "Cordwood masonry can be considered a safe material to work with with respect to fire resistance. The fire test showed that the wall is able to withstand a blast flame for over five hours. A practical fire resistance rating for the purposes of the National Building Code of Canada assembly comparison system of at least one and a half hours is recommended."

Cordwood Wall Thickness

Cordwood masonry is used for large and small houses in a variety of climates, for outbuildings such as garages, garden sheds, and saunas, and even as decorative interior walls, so wall thickness varies greatly.

Small outbuildings. Built within a stout timber frame, such as 8-by-8-inch timbers for posts and beams, an 8-inch cordwood wall works very well in any climate. Walls thinner than 8 inches make for very fussy building — I call it "cordwood needlepoint" — and any thicker is simply unnecessary. Even a sauna, typically heated to 165 degrees F or more, is a small enough building that it can be fully heated (Finns say *seasoned*) in just two hours with a woodstove, even on sub-zero winter days. You can build the walls 10 or 12 inches thick if you like, but there is little upside from a practical standpoint, and you will need 25 to 50 percent more wood and other materials. Eight inches works for a garage, too, even if it needs to occasionally be heated. A regularly heated large workspace should have thicker walls — say 12 or 16 inches, unless it is small, where, again, 8 inches is fine.

Houses. From, say, 600 square feet and up, you will want to follow energy code for your area. Indeed, you may have to. In New York, where we live, this means an R-value (insulation value) of R-19 for the walls. R-values will be discussed more thoroughly in Chapter 4, including the thermal performance of a cordwood wall and how to determine its R-value. For the moment, suffice it to say that for houses a 16-inch cordwood wall should be considered as the minimum thickness in both warm and cold climates. (Cordwood walls are effective in reducing cooling costs, too.) In Canada, cordwood walls for houses are commonly 24-inches thick, R-30+.

Small to tiny houses. 600 square feet down to ...? Energy codes usually have a one-size-fits-all attitude about R-value requirements for habitations. This can get ridiculous when we get to "tiny houses," which have become the rage in the past few years. Say your desired actual living space is just 100 square feet, so just 10 feet square. If you put an R-19 cordwood wall of white cedar around this — that is, 16 inches thick — the footprint would be 12′8″ by 12′8″, or 160.5 square feet, so that a third of the foundation footprint is caught up in the walls! With a less favorable species, say maple or oak, you would need 24-inch walls, requiring a 14′ by 14′ (196 square feet) footprint, 68 percent larger than the desired living space. With even a 16-inch overhang all around — recommended for cordwood — the roof area would be 16′8″ square, or almost 278 square feet for a 100 square foot living space! Ridiculous? Well, as they would say in cordwood's heartland — Wisconsin — "You betcha!"

The reality is that it is a challenge to heat a "code"-insulated small building without *over* heating it. We can heat a small sauna to 165 degrees in two hours with 8-inch cordwood walls, just R-8, even on the coldest winter's day.

Chapter 4

Building Science Notes

Two areas of concern often come up in discussions about cordwood masonry: thermal performance and moisture management.

Thermal Performance

Thermal performance involves *heat loss* (R-value, something building codes always specify) and *thermal mass* (something given short shrift in the building codes).

The only authoritative testing on the R-value of cordwood masonry was conducted by Dr. Kris J. Dick and Luke Chaput during the winter of 2004–2005, based on thermal sensors placed within a 24-inch-thick wall at the University of Manitoba. A paper reporting on their findings appears in *Cordwood and the Code* (see Resources).

The authors' Summary/Conclusion says, in part: "Based on approximately three months of mid-winter temperature data, the wall was determined to have an RSI Value of 6.23 (m^2K/W), R-35 for a 24-inch (60-centimeter) wall system."

While this is a very short summary of a long and detailed paper, the paper itself provides good evidence that can be used when trying to meet state or provincial R-value codes.

Insulation values of woods vary greatly amongst species. In engineering manuals, R-values for woods are generally given through side grain — as in the left-hand column of Table 4.1 — which might be appropriate for a horizontal log cabin, but they are less useful with cordwood. Some cordwood builders use these R-values for their wall calculations, but the reality is that heat transfers more readily through end grain than side grain, so an adjustment must be made. Some have suggested that end grain R-values are only 40 percent of side grain values, but I think this is unduly pessimistic. The consensus of research that I have been able to find suggests that if we use a value of two-thirds (66.7 percent) for the end grain, we will be much closer to the truth than using side grain values or the pessimistic 40 percent figure. Table 4.1 reflects side grain R-values in the left column and "closer-to-reality" values in the right-hand column for a representative sample of woods. Not surprisingly, the lighter, airier woods have better insulation value than the denser species.

Let's do a quick R-value calculation for a cordwood wall that is 50 percent wood and 50 percent mortar, by unit area. We'll say it is a 16-inch wall, with 4-inch inner and outer mortar joints, and 8 inches of insulation at R-3 per inch. We'll assume a white cedar wall, which, on end grain, yields an honest R-1 per inch. So, half of the wall's area has an insulation value of R-16 (16 inches at R-1 per inch). The mortared part is a little trickier. The mortar has very little insulation value per se, but engineers assign a value

Table 4.1: Comparative R-values for some representative woods

Wood Species	R/inch on Side Grain	R/inch on End Grain
White cedar	1.50	1.00
White pine, Aspen	1.32	.88
Basswood (Linden)	1.24	.83
Hemlock	1.16	.77
Douglas fir	1.06	.70
Red pine, Red cedar	1.04	.70
Southern yellow pine	.90	.60
Oak, Maple	.78	.52

of R-2 for the absorption of heat into one side of it and then its transfer out the other side. The insulation cavity is worth R-24 (8 inches at R-3 per inch.) Then you have the absorption and transfer of the heat through the outer mortar joint, another R-2. Add it all up, and we've got R-2 plus R-24 plus R-2, or a total R-28 insulation value for the mortared portion of the wall. Surprisingly, the mortared portion of the wall is much better insulated than the wooden part. Averaging the wood and mortared parts — remember that it's half wood and half mortar in this example — we get a value of R-22 for the wall taken as a whole (R-16 plus R-28 divided by 2 yields R-22).

New York code requires R-19. Our walls meet this requirement, and the house is very easy to heat in our area of almost 9,000 degree days, the same as Montreal. However, there is an added thermal benefit from our cordwood walls, not credited by the codebook. Due to the placement of a large thermal mass — the mortar — on each side of the insulated space, Earthwood's temperature stays steady. Even with no heat source, we can go away for a few weeks in the winter and the house does not freeze. Similarly, we do not require air conditioning in the summer. The massive walls act as a capacitor, and temperatures change very slowly. (It must be said, though, that 40 percent of our walls are earth-sheltered, which contributes to this benefit as well.)

Moisture Management

Moisture movement in or through a cordwood wall is a very close relative to air infiltration through the log-ends. When the wood shrinks, two kinds of infiltration can occur: through checking (shrinkage tangent to the edge of the log) and around the log-ends themselves (radial shrinking). Proper drying of the log-ends prior to building will greatly reduce this infiltration, but not eliminate it. And even with properly dried wood, infiltration through tiny gaps can double after a year or two, making the house more difficult to heat. A former student of mine from downstate New York built her cordwood home nearly 20 years ago. She used unsplit rounds for her log-ends, and they shrunk a good deal. After a year or two, she applied Log Jam chinking material (see Chapter 10) to the entire inner mortar joint of the home. She was meticulous, injecting the Log Jam into primary and even *small secondary* checks (checks are shrinkage cracks that extend all the way through the log-end). Peripheral shrinkage gaps were completely closed, of course. Upon completion, she found out that this chinking application — on the inside only — reduced her heating costs from $220 a month to $75 a month. Wow!

Jaki and I once visited a cordwood home with 16-inch cordwood walls in southern Wisconsin. If a strong west wind drove rain into the west wall of the home, moisture would sometimes make its way through the wall to the interior. But cordwood breathes wonderfully on end grain, and the wall would soon dry out again. We have seen a similar effect once or twice in extreme conditions on the west side of our Earthwood home, but, again, the wall soon dries out. What effect does this driven moisture have on the lime-treated sawdust insulated space? One time, bad flashing detailing on the small downstairs sunroom/greenhouse caused water to run down the cordwood wall. The log-ends were discolored but did not deteriorate. Some mortar cracking occurred. I decided to replace the wall as a workshop project, and found out that the sawdust had set up with the lime, then dried out, so that, now, instead of a loose-fill type of insulation, we had something more akin to a rigid bead board. But dry.

Another time, we removed a small section of cordwood wall to turn a window space into a door to the new upstairs solar room. The wall had never gotten wet, and the sawdust was the same light and fluffy insulation that it was when we had installed it 20 years earlier. I credit the natural breathability along the end grain of log-ends for the lack of moisture problems in the wall. In fact, cordwood walls seem to draw excess moisture *out* of the house. We maintain a pleasant relative humidity of 40 percent. In the sauna, when we throw water on the basalt stones on top of the woodstove, it turns instantly into steam, and it hits the bather almost like a physical slap. It's like a steam room for a couple of minutes. But, five minutes later, the stove room's atmosphere is back down to a very dry condition. We'll do this two or three times during a bath. There has been no deterioration in the cordwood masonry after 35 years of use.

With double-wall cordwood masonry, a vapor barrier is usually installed on the interior of the insulated space, following standard building practice. See Chapter 6.

Finally, Peter Robey and Blythe Tait built a large hexadecagon home in Tasmania. They plastered the interior — they call it "rendering" — and are pleased with the results. The rendering closed off air leaks. How they did it is told in Chapter 9.

Chapter 5

Material Specifications

CORDWOOD MASONRY requires properly seasoned wood of an appropriate species, as well as mortar materials (depending on the mortar or binder choice) and insulation materials.

Part 1. Wood

The ideal wood for log-ends

Log-end material should be light in weight for good insulation, and have a favorable (low) shrinkage value. A low shrinkage rating, incidentally, is also a good indication of the wood's disinclination to expand in damp or wet conditions. Shrinkage and expansion are two sides of the same coin. What goes up, comes down. What expands, shrinks. And vice-versa. Of shrinkage and expansion, though, expansion is more dangerous. There are various ways to deal with shrinkage, even years down the road. Wood expansion, however, can break up the wall, cause round and stackwall-cornered walls to tilt outward, and can uplift the girts of a timber frame, even raising posts off the foundation. Fortunately, these situations are rare, but my wife, Jaki, and I have experienced cordwood expansion — in our early days — and I will share with the reader various methods for avoiding the problem. And, I'll tell how to solve shrinkage problems.

Rot in wood

Rot resistance is not a big factor in choosing wood species. Wood rot is caused by fungi, which need nutrients, air, and constant moisture to propagate. With a cordwood wall, only the first two requirements are present, not the third. Because log-ends are constantly breathing along end grain, moisture is never trapped. Nevertheless, here the five things to do to prevent wood rot:

1. Keep the cordwood masonry elevated at least 4 inches off the ground on a good concrete, stone, or block foundation. In wet climates, up this to 8 inches or a foot.
2. Use a good roof overhang all around the building: 16 to 24 inches.
3. Don't allow adjacent log-ends to touch each other or a framing timber. Moisture can wick in or get trapped there, increasing the risk of fungal growth.
4. Build only with log-ends that are sound in the first place. Reject wood with any existing rot.
5. Debark the wood. Insects love to get between bark and the outer layers of the wood.

Caveat: The only wood species that I have seen used successfully with the bark left on were basswood (linden) in the East, and lodgepole pine in the West. The bark of these mature trees is hard and woody, like the wood itself, and isn't inclined to separate when the wood dries.

Preferred species

In general, light and airy woods are preferred over heavy and dense woods. They have a better R-value (insulation rating) and are less prone to expansion and shrinkage. Broadly speaking, softwood species (coniferous or "evergreen") are preferable to hardwoods, which are deciduous, losing their leaves in the winter. However, at our workshops, I avoid using the terms softwood and hardwood because some coniferous evergreens can be quite dense, heavy, and hard — such as

Southern yellow pine — while some deciduous so-called hardwoods, like cottonwood and quaking aspen, can be quite light and airy when fully dry. Wood characteristics are the key; we pass a couple of dozen different samples around the classroom so students can heft the wood, feel the hardness with their fingernails, and observe the size of the annual growth rings. Tightly packed growth rings usually mean a slower growing, harder wood — oaks and maples are

Table 5.1

Shrinkage Values of Domestic Woods

Shrinkage from green to ovendry moisture content expressed as a percentage of the green dimension.

Hardwoods

Species	Radial (Percent)	Tangential (Percent)	Volumetric (Percent)
Alder, red	4.4	7.3	12.6
Ash:			
Black	5.0	7.8	15.2
Blue	3.9	6.5	11.7
Green	4.6	7.1	12.5
Oregon	4.1	8.1	13.2
Pumpkin	3.7	6.3	12.0
White	4.9	7.8	13.3
Aspen:			
Bigtooth	3.3	7.9	11.8
Quaking	3.5	6.7	11.5
Basswood, American	6.6	9.3	15.8
Beech, American	5.5	11.9	17.2
Birch:			
Alaska paper	6.5	9.9	16.7
Gray	5.2	--	14.7
Paper	6.3	8.6	16.2
River	4.7	9.2	13.5
Sweet	6.5	9.0	15.6
Yellow	7.3	9.5	16.8
Buckeye, yellow	3.6	8.1	12.5
Butternut	3.4	6.4	10.6
Cherry, black	3.7	7.1	11.5
Chestnut, American	3.4	6.7	11.6
Cottonwood:			
Balsam poplar	3.0	7.1	10.5
Black	3.6	8.6	12.4
Eastern	3.9	9.2	13.9
Elm:			
American	4.2	9.5	14.6
Cedar	4.7	10.2	15.4
Rock	4.8	8.1	14.9
Slippery	4.9	8.9	13.8
Winged	5.3	11.6	17.7
Hackberry	4.8	8.9	13.8
Hickory, Pecan	4.9	8.9	13.6
Persimmon common	7.9	11.2	19.1
Sassafras	4.0	6.2	10.3
Hickory, True			
Mockernut	7.7	11.0	17.8
Pignut	7.2	11.5	17.9
Shagbark	7.0	10.5	16.7
Shellbark	7.6	12.6	19.2
Holly, American	4.8	9.9	16.9
Honeylocust	4.2	6.6	10.8
Locust, black	4.6	7.2	10.2
Madrone, Pacific	5.6	12.4	18.1
Magnolia:			
Cucumber tree	5.2	8.8	13.6
Southern	5.4	6.6	12.3
Sweetbay	4.7	8.3	12.9
Maple:			
Bigleaf	3.7	7.1	11.6
Black	4.8	9.3	14.0
Red	4.0	8.2	12.6
Silver	3.0	7.2	12.0
Striped	3.2	8.6	12.3
Sugar	4.8	9.9	14.7
Oak, red:			
Black	4.4	11.1	15.1
Laurel	4.0	9.9	19.0
Northern red	4.0	8.6	13.7
Pin	4.3	9.5	14.5
Scarlet	4.4	10.8	14.7
Southern red	4.7	11.3	16.1
Water	4.4	9.8	16.1
Willow	5.0	9.6	18.9
Oak, White:			
Bur	4.4	8.8	12.7
Chestnut	5.3	10.8	16.4
Live	6.6	9.5	14.7
Overcup	5.3	12.7	16.0
Post	5.4	9.8	16.2
Swamp chestnut	5.2	10.8	16.4
White	5.6	10.5	16.3
Tupelo:			
Black	5.1	8.7	14.4
Water	4.2	7.6	12.5

examples — while large annual growth rings usually translate into lighter, airier woods with better insulation value and a lower shrinkage coefficient. There are always exceptions, though. For example, hemlock, which can weigh 110 percent more in its green state than when it is fully air dry, suffers from up to 11 percent shrinkage as it seasons from green to dry. This doesn't mean you can't use it; but you need to be aware of what it is likely to do and plan the wood seasoning and build schedules accordingly.

Look at Table 5.1: "Shrinkage Values of Domestic Woods." While *domestic,* refers to the United States, you will find most common Canadian species listed as well. Note that three kinds of shrinkage are listed in the chart: radial, tangential, and volumetric. Radial shrinkage is shrinkage following the radius of a round

Table 5.1 ***continued***

Shrinkage Values of Domestic Woods							
	Shrinkage from green to ovendry moisture content expressed as a percentage of the green dimension.				Shrinkage from green to ovendry moisture content expressed as a percentage of the green dimension.		
Species	**Radial**	**Tangential** *Percent*	**Volumetric**	**Species**	**Radial**	**Tangential** *Percent*	**Volumetric**
Hardwoods continued							
Sweetgum	5.3	10.2	15.8	Walnut, black	5.5	7.8	12.8
Sycamore, American	5.0	8.4	14.1	Willow, black	3.3	8.7	13.9
Tanoak	4.9	11.7	17.3	Yellow-poplar	4.6	8.2	12.7
Softwoods							
Baldcypress	3.8	6.2	10.5	Larch, western	4.5	9.1	14.0
Cedar:				Pine:			
Alaska-	2.8	6.0	9.2	Eastern white	2.1	6.1	8.2
Atlantic white-	2.9	5.4	8.8	Jack	3.7	6.6	10.3
Eastern redcedar	3.1	4.7	7.8	Loblolly	4.8	7.4	12.3
Incense-	3.3	5.2	7.7	Lodgepole	4.3	6.7	11.1
Northern white-	2.2	4.9	7.2	Longleaf	5.1	7.5	12.2
Port-Orford-	4.6	6.9	10.1	Pitch	4.0	7.1	10.9
Western redcedar	2.4	5.0	6.8	Pond	5.1	7.1	11.2
Douglas-fir:				Ponderosa	3.9	6.2	9.7
Coast	4.8	7.6	12.4	Red	3.8	7.2	11.3
Interior north	3.8	6.9	10.7	Shortleaf	4.6	7.7	12.3
Interior west	4.8	7.5	11.8	Slash	5.4	7.6	12.1
Fir:				Sugar	2.9	5.6	7.9
Balsam	2.9	6.9	11.2	Virginia	4.2	7.2	11.9
California red	4.5	7.9	11.4	Western white	4.1	7.4	11.8
Grand	3.4	7.5	11.0	Redwood:			
Noble	4.3	8.3	12.4	Old-growth	2.6	4.4	6.8
Pacific silver	4.4	9.2	13.0	Young-growth	2.2	4.9	7.0
Subalpine	2.6	7.4	9.4	Spruce:			
White	3.3	7.0	9.8	Black	4.1	6.8	11.3
Hemlock:				Engelmann	3.8	7.1	11.0
Eastern	3.0	6.8	9.7	Red	3.8	7.8	11.8
Mountain	4.4	7.1	11.1	Sitka	4.3	7.5	11.5
Western	4.2	7.8	12.4	Tamarack	3.7	7.4	13.6

Coast Douglas-fir is defined as Douglas-fir growing in the States of Oregon and Washington west of the summit of the Cascade Mountains. Interior West includes the State of California and all counties of Oregon and Washington east of but adjacent to the Cascade summit. Interior North includes the remainder of Oregon and Washington and the States of Idaho, Montana and Wyoming.

log-end. You can also think of it as shrinkage across the annual growth rings. Tangential shrinkage follows the circumference of a round log-end, that is: tangent to the cylinder. Volumetric shrinkage is as the term implies: the total shrinkage of wood, in volume. You will see that the combination of radial and tangential shrinkage is close to the volumetric shrinkage, but rarely exactly so.

Tangential shrinkage in a round log-end manifests itself by checks which open tangent to the annual growth rings, that is, the checks follow a line from the center of the log to the exterior. As the wood dries, these checks make their way through the log-end and expand or open wider, shrinking in directions tangent to the log-end's perimeter. See Figures 5.1 and 5.2. On a round log-end, one of these checks — which we call the primary check — will finally win out over its fellows, and, when it does so, it breaks through from one end to the other with a loud audible cracking sound. We — and other cordwood builders — have reported this sound occurring in completed cordwood walls, even a year or two after occupying the home. Once the primary check breaks though, the log-end can now easily shrink the way it wants to — tangentially — and, as a result the several other checks which formed as the log-end began to dry, actually close up, because the wood is shrinking at the line of least resistance: the primary check.

With a round log-end, this primary check can get quite large — or not so large, depending on species and drying rate. See Figures 5.3 and 5.4. With wood that is split soon after cutting it into log-end length, hardly any checking will

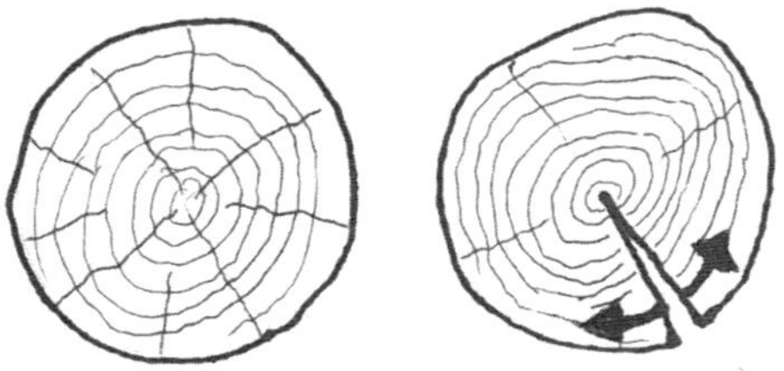

5.1: *Left: Several checks form as the log-end begins to dry. Right: One of them breaks through and further shrinking takes place tangentially in two directions, as per the arrows.*

5.2: *Several checks began to form during the initial drying, but one wins out, becoming the primary check; this can be filled with clear caulking down the road.*

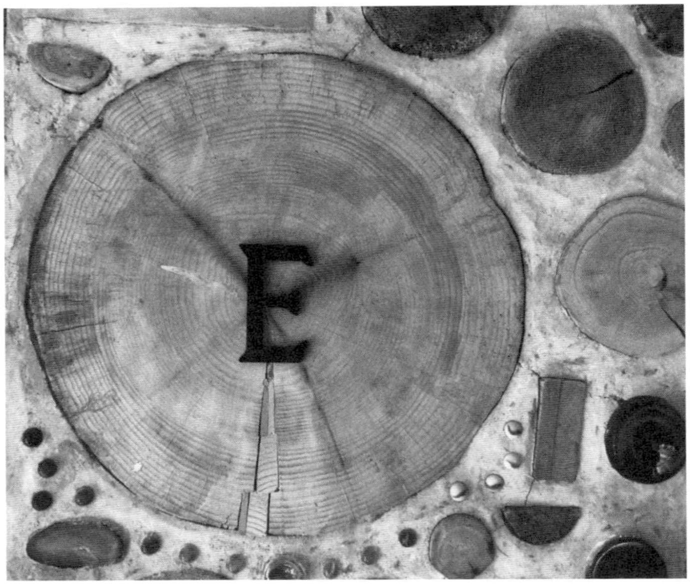

5.3: *The primary check of this large 17-inch diameter log-end shrunk so much after it was laid up that I actually filled the space with fitted wood pieces. Then it shrunk a little more. Note that radial shrinkage, around the perimeter, was very little and fixed easily with a bead of clear caulking.*

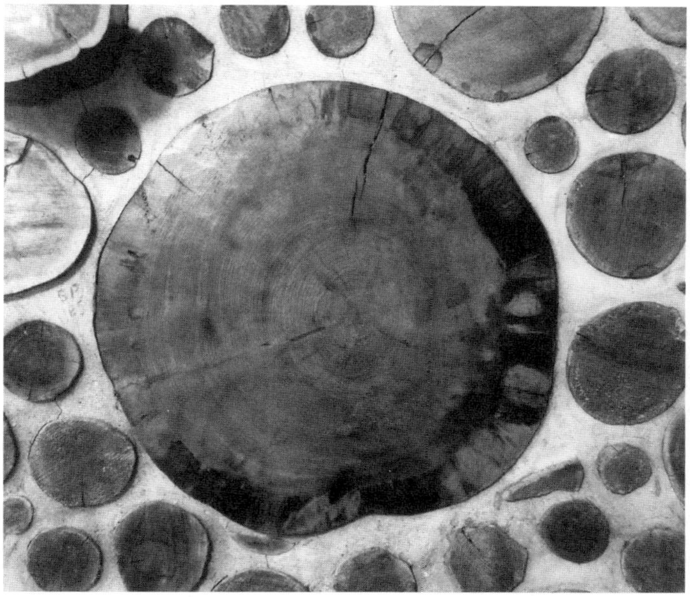

5.4: *This 27-inch diameter (12-inch thick) red pine special-feature log-end at our Mushwood Cottage has suffered from very little shrinkage, considering its size.*

5.5: *This split rock maple log-end has no discernible checking, even after 30 years in the classroom. Its end grain, however, feels like hard granite to a fingernail test.*

occur, because tangential shrinkage can begin immediately. See Figure 5.5. An exception to the "one check wins out" rule is shown in Figure 5.6; at workshops in Australia, we sometimes observed three or four substantial checks on eucalyptus log-ends. There may be other species that perform in this way that I am not aware of, but in workshops throughout North America, a primary check on round log-ends is the norm. Rarely, we will see both a primary and a secondary check on a large log-end, even white cedar, our favorite choice.

Splitting a log-end doesn't eliminate tangential shrinkage; it only stops checking. The shrinkage is still there, following the same line from the center of the log to its exterior, as seen at the top edge of the split log-end shown in Figure 5.7.

Timbers dry about ten times faster through end grain than through side grain. New England shipbuilders knew this very well, and, to avoid great checks in their ships' masts and booms, they would wax-seal the ends of long logs,

5.6: *Eucalyptus does not seem to follow the "primary check" rule. It also doesn't split very well. In fairness, there are more than a hundred varieties, and some might perform differently from others.*

forcing them to dry slowly through side grain, greatly reducing checking. Wood carvers and turners use another trick. Not wanting checks in their bowls or duck decoys, they will turn or carve large quarter-sawn logs, eliminating the primary check formed by tangential shrinkage.

5.7: *Radial shrinkage in this split hardwood log-end is seen clearly on the left outer edge. The top edge shrinking is tangential shrinkage.*

5.8: *This hardwood log-end shows radial shrinkage around its perimeter, as well as tangential shrinkage in the form of checking.*

Radial shrinkage follows the radii of a round log-end; that is: perpendicular to the annual growth rings. This is true even if a round log-end has been split. See Figures 5.7 and 5.8. The shrinkage takes place from the circumference toward the log's center, manifesting itself as a shrinkage gap all around the log-end. Shrinkage is a proportional phenomenon. All else being equal (species, drying time, length of log-end) a 20-inch diameter log-end would have perimeter shrinkage gaps about twice as thick as a 10-inch log-end, or four times as great as a 5-inch log-end. However, the total wood shrinkage occurring in the wall will be the same, whether or not large log-ends are split into smaller pieces.

As a general rule, tangential shrinkage is about twice as great as radial shrinkage.

Volumetric shrinkage can be thought of as a combination of tangential and radial shrinkage. With split wood, the shrinkage takes place all around the log-end. *Longitudinal shrinkage* (along a log's length) is miniscule and does not affect cordwood masonry at all.

Shrinkage — and its brother, expansion — are potential problems with cordwood masonry, which begs the question, "How long should I dry the log-ends?" This sounds like a reasonable query, but it is something like "How high is up?" First, we need to know how long these log-ends are going to be, which depends on the thickness of the wall, already discussed in Chapter 3.

Seasoning cordwood

Students often ask: "I've had these logs lying around for three years now. They must be pretty dry by now, right?" Rob: "What length are these logs?" Student: "Oh, 8 or 10 feet."

I explain that wood takes about a year per inch of thickness to fully dry on side grain, but that it dries ten times faster through end grain. "Only the two ends of your long logs have done any significant drying," I tell them. "You really need to get the long logs cut into log-end length before any real drying can take place." This is not exactly what they want to hear. But the follow-up question is more to the point, to wit: "Then how long do I need to dry them in the log-end length?" The answer to this one can take 30 to 40 minutes of classroom discussion, because a lot of considerations must be factored in: the thickness of desired wall, the kind of wood, whether it is split or round, and what the drying conditions are.

Thickness of wall. We'll presume you know how thick you want your cordwood walls, based partly on the discussion above. For the lighter and airier woods (generally, those in Table 5.1 with the lowest shrinkage rates), a full year's air-drying — as described below — is advised for exterior walls made with 12- to 16-inch log-ends. If this rule is followed, you should not have to do a total mortar joint chinking in future. A few of the larger log-ends may require some clear caulking around their perimeters or in the primary checks, but most of them will be fine. Jaki and I have been building with cordwood for 40 years and have never had to do full mortar-joint chinking, but we have attended to large special-feature log-ends with clear caulking. Remember that shrinking is proportional: big log-ends, big shrinkage gaps; small log-ends, small shrinkage gaps (although the total net air infiltration is the same.) With 8-inch log-ends and favorable drying conditions (discussed below), we have stacked our 8-inch cordwood to dry in April and used it without a problem as early as July, and certainly by September.

Kind of wood. With the more favorable (less dense) species, over-drying is not a problem — providing basic building precautions are taken, as described in Chapter 8. Following the recommendations in the paragraph immediately above is pretty safe. However, if you are forced to use dense, heavy woods, such as oak, beech, sugar maple, red cedar, or Southern yellow pine, then there is the very real danger of over-drying the wood. If too-dry dense woods take on moisture, such as from a driving rain, they will very likely swell; this can cause structural problems, breaking up mortar joints and causing wall tilt-out (in the case of load-bearing cordwood walls) or the uplifting of a timber frame (where cordwood is used as infilling). This dire warning does not mean that you can't build with these woods, but it does mean that you must take precautions as described in the Sidebar here.

On a related note: yes, wood species can be mixed, as long as they are good species choices and a consistency of texture and style is maintained.

If Hardwood Is All You've Got...

Many cordwood homes have been built successfully with hardwood. The following precautions will greatly minimize the danger of expansion and structural damage:

1. Don't use overly dry dense woods. Limit the drying time to a month with splits, two months with rounds. Even with a year's seasoning, the log-ends will probably still shrink in the wall after another year or two. There are things that can be done about shrinkage (discussed in Chapter 10), but wood *expansion* is a disaster.
2. For weather protection, build within a timber frame only after the roof is on. And incorporate a good overhang all around the building.
3. Hardwood log-ends are not recommended for round or stackwall styles of cordwood masonry.
4. Don't build on an uncovered slab where water can collect against the cordwood.
5. Keep hardwood log-ends off the foundation. Use more favorable woods down low, or use insulated blocks or stone masonry for the first foot of the wall.

Split or round. Beautiful, successful cordwood walls have been built from all rounds, all splits, and a pleasing balance of splits and rounds. We've done all three. (See Sidebar, "Split Versus Round.") As for drying time, split wood dries quite a bit faster than nonsplit wood, as wood burners know from experience. How much faster is a guess ... but *my* guess is that a round log-end split into four equal quarters will probably dry 30 to 60 percent faster than in its

Split Versus Round

The three main reasons for splitting wood are to reduce the drying time, eliminate the large *primary check* seen in rounds, and reduce the size of shrinkage gaps. Shrinkage is proportional, so the smaller the log-end, the smaller the shrinkage gap between wood and mortar. But the total air infiltration in the wall due to shrinkage is the same. Also, smaller pieces require more handling of materials, and mixing more mortar, too.

Jaki and I have built cordwood walls with all split wood, all rounds, and a mixture of splits and rounds. We like the ability to keep a constant mortar joint by using a variety of splits, but, since the turn of the century, we have been using cedar rounds almost exclusively. We like the appearance of the various sizes of rounds, the ease of pointing, and the elimination of the splitting step. We are careful to keep the primary checks oriented down, though, ☛

5.9: *All rounds. All splits. Splits with featured rounds.*

5.10: *All rounds (left). All splits (center). Mixed splits and rounds (right).*

fully round state. Longer log-ends would tend toward the 30 percent faster figure, shorter log-ends toward the 60 percent faster figure. This will also vary with species. For example, quaking aspen logs are very heavy in their green state; they take much longer to dry as rounds than as splits.

Drying conditions. As you might expect, wood dries faster in drier climates than in damp, coastal, or otherwise humid climates. And

somewhere between 4 o'clock and 8 o'clock, so that rainwater doesn't collect in the checks.

Split woods can have ragged or irregular ends due to the splitting process, but they don't usually have any tangential checking. Some woods, like red pine, grow spirally in the forest, so the axe or splitter can take a 20-degree turn — or more — as it makes its way through. Twisted log-ends don't play well with the other log-ends.

Mixing splits and rounds can be very effective, if care is taken to maintain a consistency of texture and style, as in Figure 5.11.

Having said all that, if you have a strong preference, go with it. All three styles will work. The important thing is to maintain a visual or textural consistency, which means making a conscious effort to deplete the various sizes and shapes of your log-ends at a steady rate.

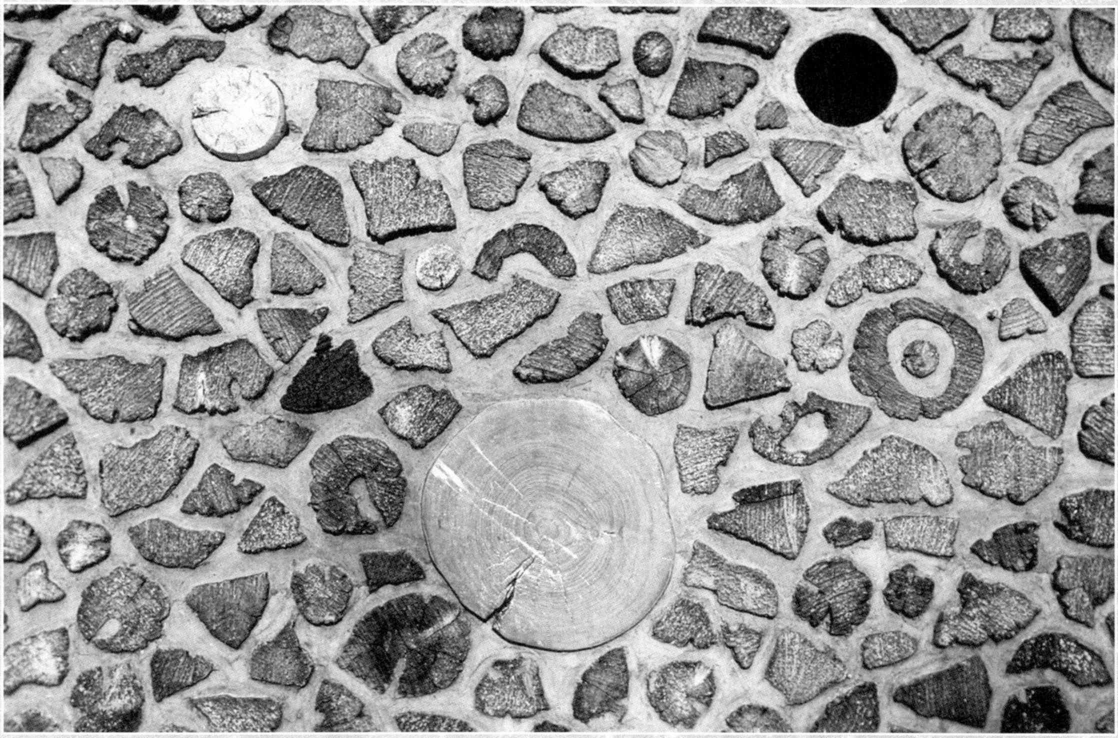

5.11: *Our Log End Sauna walls were mixed rounds and old split cedar fence rails. The large log-end was a varnished elm. It stayed nice for over 30 years on the inside, but the sun's UV rays wore the varnish off on the exterior within a year or two. The two rounds at the top had handles on the inside and could be removed for ventilation.*

microclimates need to be understood. One side of a mountain range could be radically different from the other side. Drying conditions in our moderate rainfall area vary greatly from year to year. The year we built our new sunroom at Earthwood, we stacked our 8-inch and 16-inch cedar log-ends for drying in April; we were able to use the 8-inchers in July and the 16-inchers in September. We got lucky with a relatively hot, dry spring and summer. Normally, the 16-inch log-ends would need to be dried for a year before use.

How much wood is needed?

Calculating the quantity of cordwood (log-ends) required for the job is fairly straightforward — once you know how much of your external walls will actually be cordwood masonry. For this, you will need elevation plans, the drawings that show you what the sides of the building look like. In a typical rectilinear design, this means four separate elevations. Our Stoneview Guesthouse is an octagon, so I had to look at and calculate cordwood area for eight different elevation drawings, (although three of them were identical, as were two others). So, let's say you build with a timber frame where it is 7 feet from the foundation to the underside of the girts. (A girt is the horizontal member that connects the tops of the posts.) In this case, multiply 7 feet times the length of the panel, the space between posts. Now, subtract the cross-sectional area of all windows and doors (including their thick frames). Add up the square footage for all the panels. If you have cordwood masonry earmarked for gable ends of the building, include that, too. This is the gross square footage of cordwood masonry in the building. You can work from the outside elevations because the inside of the panels will have exactly the same area of cordwood.

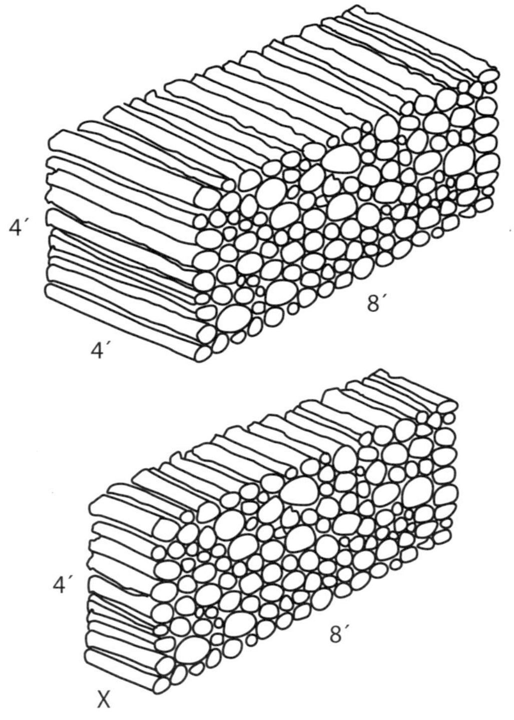

5.12: *Above is a full cord, below is a face cord, where "X" is whatever length of wood the seller is supplying: 12 inches, 16 inches, etc.*

Okay, that was the hard part. The math is pretty easy. We work in a unit called a *face cord,* defined (depending on where you live) as a *stack, rank,* or *rick* of wood measuring 4 feet high and 8 feet long with a width of whatever length the logs or log-ends are cut to. A true, real or "full" cord is actually 4 feet high, 8 feet long, and 4 feet wide — 128 cubic feet. But, usually, when you buy firewood, the wood merchant is selling a cord which is 4 feet by 8 feet by — typically for firewood — 16 inches. Or he might be selling "foot-wood," where the width is only 12 inches. Know what you're buying. Some merchants still sell uncut "full cords" of 128 cubic feet, but this is usually limited to wood being sold for pulp. For these calculations, your "face cord" will be 4 feet by 8 feet by *X,* where *X* is the width of your cordwood wall, which is the width of your log-ends. The formula works for face cords no matter if *X* is 8, 12, 16, 24 inches, or whatever.

Step 1. Figure the gross square footage (square feet) of actual cordwood masonry for your plan, as discussed above. For this example, we are going to assume a total cordwood area of 640 square feet, maybe a small cottage. But remember that the actual square footage must be calculated from your elevation plans.

Step 2. Divide the cordwood square footage (square footage) by 32, because there are 32 square feet on the side of a face cord. In our example, we divide 640 square feet by 32, yielding 20 face cords.

Step 3. Because the cordwood masonry has a lot of mortar between the log-ends — typically a 60/40 ratio, wood to mortar — we can safely take 80 percent of the face cords calculated in Step 2. So, 20 face cords times .80 equals 16 face cords actually needed. I use an 80 percent factor — instead of 60 percent — for three reasons. (1) The wood stacked in the face cord has a lot of air space between the log-ends, which is like mortar space but obviously not as much. (2) Maybe you use thinner mortar joints than we do, say 70/30 wood/mortar ratio. (3) This formula allows you to reject log-ends that don't quite meet quality standard: pieces that are twisted, deteriorating, too short, or just plain ugly. Worst case scenario? You have a few log-ends left over. As you are probably going to heat with wood anyway, they will not go to waste. But running a few log-ends *short* is a bummer. I have been using this formula — and teaching it — for years and have never run short of log-ends, and have had no complaints from other builders who use it.

Part 2: Mortar Materials

Cement and lime

Based upon the hundreds of cordwood buildings I am familiar with in North America, I would estimate that 90 percent of them make use of cement-and-lime-based mortars, and these will be the focus of this section. But there are other choices, such as lime putty mortar, papercrete, hempcrete, and cob between log-ends, and they will be discussed in *Part 3: Green Mortar Options.*

Cement mortars can be made with either Portland cement or masonry cement. We have used both. Our first homestead, Log End, used Portland and lime mortar. Earthwood, our home of 35 years, was built with masonry cement. More recently, at our Mushwood summer cottage and at the various Earthwood guesthouses, we have gone back to the Portland-based mix. There is also something called "mortar mix." Here's the difference between the products.

Portland cement. *Type I* or *Type II* (air-entrained), is full-strength, relatively fast-setting cement. It is a fairly recent product, about 200 years old. Lime-based mortars had been used in the old and new worlds for thousands of years before Portland's development. Bricklayer Joseph Aspdin, in England, patented Portland cement on October 21, 1824. In 1872, David O. Saylor made the first cement in the United States, in eastern Pennsylvania, but it was not until about 1900 that Portland cement became as commonly used as lime mortars. A full cubic-foot bag of Portland cement always weighs 94 pounds.

Masonry cement. This has characteristics of both Portland and lime and varies according to type, but you can always be sure of the strength of types *M, N,* or *S.* Masonry cement can vary from 70 to 80 pounds, depending on type.

No matter which cement you choose, make sure it is fresh. If the bags are rock hard, they're no good. If you open a bag and find that part of it is nice powder and part has set (as if it got wet), return the bag to the supplier for a replacement. Some suppliers have a good turnover and their cement is fresh. Others have cement bags that have been lying around for years.

Mortar mix is a pre-mixed mortar, with a ratio of about three parts sand to one part masonry cement. Just add water. If you substitute mortar mix for the masonry cement in the Masonry Cement Mix formula given below, as at least one builder has done (inadvertently), the result is untenably weak mortar. However, mortar mix alone can be used in combination with soaked sawdust or a commercial *cement retarder,* both to be discussed. We tested this at one of our guesthouses. Structurally, it is fine, but the mortar was darker in color, and we found that the sand component was rather coarse, making for less plasticity and rougher pointing.

Builder's lime. Both our Portland and masonry cement cordwood mortar recipes below involve a heavy dose of lime. Be sure to use *builder's lime,* also called *mason's lime.* This is *Type S hydrated lime.* You get it where masonry products are sold; it is different from the non-hydrated lime used in agriculture, which will not work as a mortar admixture. To be absolutely sure, *masonry* or *building* should be referenced on the bag. Builder's (or "mason's") lime will calcify (harden) over time, but its main purpose is to make the mortar more plastic and easier to use right out of the wheelbarrow.

Proven recipes

Over the first six years of our cordwood masonry experience, Jaki and I gradually refined a cement-based mortar mix that has worked very well for us. It makes use of saturated sawdust as a cement retarder (which delays drying). A mortar that dries slowly will shrink less (or not at all), eliminating mortar shrinkage cracks between log-ends.

Suitable sawdust. The problem is that the right kind of sawdust is not always available, and there is always a little bit of doubt about whether or not the available sawdust is right for the job. "Suitable" sawdust, in our experience, consists of the larger and less dense particles of softwood sawdust that come from a sawmill where logs are made into lumber — as opposed to what you get from sanders or planers at a cabinetmaker's shop. White cedar, white and red pine, lodgepole and ponderosa pine, spruce, and even quaking aspen sawdusts have worked well. Note that species which make good log-ends are generally the ones that work well as sawdust additives. Oak and other dense hardwood sawdusts have not been successful. The hard little cubes of oak do not seem to hold and store the moisture the way the softer, lighter softwoods do, and mortar shrinkage is the result. In fact, hardwood sawdust makes the mortar grainy, crumbly, and difficult to point smoothly. If you cannot get suitable sawdust — or are unsure — use one of the commercially available cement retarders, discussed in the Sidebar, "Commercial Cement Retarders."

We learned about "sawdust concrete" 30+ years ago from a report by — I think — the

Table 5.2: Environmental impacts of lime

Ecosystem Impacts	Embodied Energy	Carbon Footprint	Indoor Environment	Waste
Limestone is a non-renewable resource but is abundantly available. Large scale quarrying can cause habitat destruction and surface and ground water interference and contamination.	5.3 MJ/kg. Lime is processed at high temperature, in addition to quarrying, crushing, and transport energy input.	0.78 $kgCO_2e/kg$. Lime will absorb CO_2 during the curing process, but due to fuel use during processing will still be a net carbon emitter, though accurate figures are difficult to assess.	Lime-based mortar or plaster can contribute to high indoor air quality, providing natural antiseptic qualities and no toxic off gassing.	Construction: Lime can be stored or used as a soil amendment. End of life: Lime can be broken up and used as fill.

University of Pennsylvania. Slow setting was one of the characteristics. Based upon this limited reference, Jaki and I experimented with various quantities of soaked sawdust over a two-year period. The mixes listed below were the hardest, strongest, smoothest mixes that still retained the non-shrink advantage. Less sawdust than specified resulted in mortar shrinkage and cracking.

How does the soaked sawdust retard the mortar set? There are two possibilities: (1) You are introducing millions of little "sponges" — the soaked sawdust particles — each storing water in the mix. As the mortar gives up its moisture to chemical setting and evaporation, this reservoir of stored moisture replenishes the loss, slowing the set. Or, it could be: (2) Lignum, a wood chemical, is found in commercial cement retarders, sometimes as a primary ingredient. Perhaps the soaked sawdust is creating a lignum "soup" introduced into the mortar. Maybe the answer is a combination of these two possibilities.

Sand. The sand should be washed masonry — or *mason's* — sand, not the coarse-grained sand used for drainage applications. You may have to pay more for the finer-grained masonry sand (which has the texture of granulated sugar) than for the coarse stuff, but it is worth it. Besides, haulage is usually the greater cost, not the sand itself. Coarse sand yields crumbly mortar, frustrating to work with. Also, the color of the sand will affect the color of the mortar. Light-colored sand gives light-colored mortar. Dark sand, dark mortar. On Hawaii's Big Island we used black sand and got dark mortar, which was fine for that application. But keep in mind that a cordwood masonry wall is a *light sucker.* As a significant portion of the wall's area is mortar, a lighter mortar can go a long ways toward making brighter internal spaces.

Assuming suitable sawdust — and good sand — here are two mixes that have both worked well for us, one with Portland cement and one with masonry cement. The proportions given are equal parts by volume, such as rounded shovelfuls.

Portland Mix
- 9 parts sand
- 3 parts soaked sawdust
- 3 parts lime
- 2 parts Portland cement

Masonry Cement Mix
- 9 parts sand
- 3 parts soaked sawdust
- 3 parts masonry cement
- 2 parts lime

These two mixes are similar in terms of hardness, strength, workability, plasticity, and smoothness. The main difference is in color. The Portland mix tends to be very light in color, a good thing, kind of a green-gray. The masonry mix is darker and more of a blue-gray. But even these generalities can vary when different brands of cement are used, so be consistent with your purchases. The actual mixing of the mortar is treated in Chapter 8.

The sawdust mortar fingernail test

I recommend that you get a small amount of the available sawdust — a few shovelfuls — and test it in a batch of mortar — *before* getting a large quantity. Make a batch according to the recipes above and the Chapter 8 instructions. Lay up a few log-ends for practice. If the mix is working properly, you will be able to scratch it easily with your fingernail the following day. The day after that, you will still be able to scratch it, but with difficulty. By the third day, you will not be able to scratch it with your fingernail.

But watch the test panel for another few days. If the mortar is going to shrink — and crack — this will generally occur within a week of laying it up, maybe as soon as five days. If this happens, you should use a cement retarder instead of the sawdust. (But don't use both sawdust *and* a cement retarder. The mortar might never set.) For more on cement retarders, see Sidebar.

Commercial Cement Retarders

When suitable sawdust is not available for use as a cement retarder, use a commercial product made for the purpose. Most commonly, it is a liquid, available in gallon or five-gallon containers, but it can also come as a powder. The mortar mix is slightly different when using a retarder: an extra part of sand replaces the three parts of soaked sawdust. So, with Portland, the recipe is 10 sand, 2 Portland, 3 lime. With masonry cement, it is 10 sand, 3 masonry cement, 2 lime.

I have used three or four different retarders over the years, all with good success. One was Sika Plastiment, now known as SikaTard-R. Another was Daratard-17 from W.R. Grace and Company.

Sometimes it is hard to find cement retarder at your local building supply. You may be met with a blank face by the clerk. But concrete batch plants always have a large vat of cement retarder which they add to concrete when they are pouring certain jobs that require a slow set, such as bridges. Walk in to the concrete dispatch office with a plastic gallon milk bottle in one hand and a six-pack of a good micro-brewed beer in the other. Offer the beer for a gallon of their best draft cement retarder. Usually works. Once, in Asheville, North Carolina, I didn't have the beer, but offered to pay for a gallon. It took four guys in the office about a half hour to figure out that it would be $7.50. They would have been better off giving it to me free the moment I walked in.

What's in cement retarders? According to *Design and Control of Concrete Mixtures* 14th edition, by Steven H. Kosmatka, Beatrix Kerkhoff, and William C. Panarese (Portland Cement Association), the ingredients are usually lignin, borax, sugars, Tartaric acid, and salts. You may want to check the ingredients of the ones you find locally.

Other cement retarders are listed in Resources. Their listing does not constitute an endorsement, and you may find others at local suppliers, such as Home Depot or Lowe's.

Part 3: Green Mortar Options

Some green builders object to the use of Portland cement, and I can't argue with them. The manufacturing of Portland cement — the key ingredient in concrete — is an extremely energy-intensive process, putting at least five percent of the human-induced carbon dioxide into the atmosphere. And yet, so many "green" or sustainable builders put their structures on — yep — *concrete foundations.* In my *Cordwood Building* book (New Society, 2016), the entire 1,700-word Chapter 5 addresses the question, "Is Cordwood Green?," with extensive commentary on its sustainability, low embodied energy, energy efficiency, healthy use of materials, and leaving a low impact on the planet. I will not rehash those many words here — if you've gotten this far, it's a moot point — but I will say that I stand by my conclusion from that chapter: "Is Cordwood Masonry green? Well, in this author's admittedly biased opinion, it compares very favorably with any other building method."

Nevertheless, there *are* alternatives to Portland-based mortars.

Lime Putty Mortar

I have learned a great deal about lime putty mortar (LPM) from friend and cordwood builder Bruce Kilgore, who kindly checked over my narrative below. Lime mortar has come back in vogue with the modern natural building movement, but it has been around a long time, at least since Minoan Crete about 3,000 years ago. Romans, building on the work of the Greeks, greatly improved lime mortar. The writer and engineer Vitruvius famously said of lime mortar, "After slaking it, mix your mortar, if using pit sand, in the proportion of three parts of sand

Table 5.3: Environmental impacts of portland cement

Ecosystem Impacts	Embodied Energy	Carbon Footprint	Indoor Environment	Waste
Portland cement is made from limestone, a non-renewable resource that is abundantly available. Large scale quarrying can cause habitat destruction and surface and ground water interference and contamination.	5.5 MJ/kg. Portland cement is processed at high temperature, in addition to quarrying and crushing energy input.	0.95 $kgCO_2e/kg$	Cured Portland cement is benign and will have little impact on indoor air quality.	Waste is benign and can be used as fill.

Sources: Inventory of Carbon & Energy Database (ICE) http://www.circularecology.com/embodied-energy-and-carbon-footprint-database.html#.WO1KXGe1upo and Chris Magwood, *Making Better Buildings* and *Essential Hempcrete Construction*.
Note that when calculating environmental impacts of mortar, a typical mortar mix for cordwood masonry is 9 parts sand, 3 parts lime, 2 parts Portland cement, and 3 parts soaked sawdust, or 10 parts sand, 3 parts lime, 2 parts Portland cement, and 3 oz. of commercial cement retarder.

to one of lime." And why is LPM used today? Well...

Advantages of LPM

- LPM requires only Type S hydrated lime, sand, and water.
- It is very light in color, a plus with light-absorbing cordwood masonry walls.
- Mortar pointing can often be done the next day, if necessary.
- It is more environmentally friendly than cement mortar, having much less embodied energy in the manufacturing process.
- LPM is pleasing to work with; it is very cohesive and plastic.
- Calcification can close up small gaps and cracks in LPM over time.
- Lime-based products, such as mortar and plaster, offer superior breathability to cement-based products.

Disadvantages of LPM

- Type S hydrated lime may be hard to come by in certain parts of the country.
- Lime varies greatly in price.
- The lime putty should be made a minimum of three days in advance, although Bruce Kilgore recommends five days. (For a large project, like a house, it is good to make several large batches at a time. It will keep, as long as it is covered with plastic.)
- LPM is subject to frost damage for a longer period of time than other mortars. Special care must be taken using LPM if temperatures of 30 degrees Fahrenheit or less are likely within a two-week period after laying up the wall. With cement mortars you need only be concerned for two *days*, not weeks.
- Full strength — or nearly full strength — may take a month to achieve. Portland cement is hard and strong in three days. This is not a huge drawback unless you want to build a large, load-bearing cordwood wall fast.
- Bruce, with over a thousand batches of LPM behind him, notes that lime mortar has "a more narrow range of forgiveness" than Portland-based mortar. He emphasizes the need to pay attention to detail.

Making Lime Putty and Lime Putty Mortar

Success with LPM is a matter of minimizing variables. Careful quality control has produced good results, whereas there have been failures where variables are not kept within best practices.

Get the Right Stuff

The first and most important variable is to use the right lime, which is *dry hydrated Type S lime.* It comes in nominal 50-pound bags. Building or masonry use should be referenced on the bag. Also, be wary of lime which is more than six months old or has been improperly stored. A broken bag should not be used. Bruce is a stickler for getting his lime as fresh as possible. While trying to perfect LPM, he discovered that bags of lime can vary by five pounds or more. "That was the *aha* moment," he says. He began to weigh everything — lime bags, lime putty, sand, even water — which led to diminished variables and high-quality LPM.

Accuracy in Measure

Accuracy is important. For making lime putty, Bruce weighs three bags of lime in pounds. (Use a respiration mask when working with lime.)

5.13: *After measuring the right amounts of lime and water into half of a 55-gallon plastic drum, everything is homogenized with a paddle drill.*

Then he measures out a weight of water which is three-quarters the weight of the lime. If three bags weigh 160 pounds, for example, he mixes in 120 pounds of water to make the lime putty. To this water, he mixes in a cup of inexpensive dishwashing liquid, which acts as a *surfactant,* making water wetter and the lime putty easier and faster to mix. He starts with about a third of the water in the bottom of a plastic vessel made from half of a 55-gallon plastic drum. He adds a bag of lime and lets it sink in, then homogenizes it with a paddle mixer attached to a strong (one-half horse power) electric drill, as per Figure 5.13. Then he adds another third of the measured water, being careful to pour the water onto a mason's trowel "floating" on the first third so as not to upset the stuff below. Then he adds another bag of lime and repeats the procedure, and then once more. After all the layers are added, he homogenizes the entire mix one more time with the paddle drill. The lime putty is allowed to hydrate five days.

Cover the top of the container with plastic. The lime putty is ready to use after five days, but it will only get better with age. Bruce has used it after weeks, even months, with excellent results. It is said that in olden times, a keg of lime putty was a good and valuable addition to a bride's wedding dowry, along with a cow, goat, etc.

Sand: A Critical Variable

Use the same fine masonry sand used with Portland-based cordwood mortar. Theoretically, coarser sand should work, but using coarse sand at a Colorado workshop resulted in a less cohesive (plastic) mortar and, ultimately, the mortar became crumbly (although this may have had more to do with the extreme drying conditions in the clear dry air at 8,200 feet).

It is of great importance that the sand be kept dry. If it is too wet, it will not be possible to make a stiff enough mix to use with cordwood masonry. So, get dry sand — and keep it dry — by keeping it well-covered. You can add water to a mix, if needed, but you can't take it out if the mortar is too wet before any water is added. You cannot add dry lime to the mix in order to stiffen it up. All lime used must be previously hydrated.

Mixing the Lime Putty Mortar

Bruce, a stickler for detail and consistency, measures each wheelbarrow load (batch) by weight: 75 pounds of sand and 28 pounds of lime putty, already prepared as described above. He mixes the ingredients first with a hoe until the mix is a consistent color throughout. Then he kneads the mortar with his gloved hands against one end of the wheelbarrowl, much as pizza dough is kneaded. He kneads the mortar toward one end of the wheelbarrow, then the other. If it passes the stiffness tests (see below), it is finished.

Organizing the Mixing Area

Bruce takes great care to organize his mixing area for efficiency and consistency of mixing. He uses six lime putty drums, each one being a half of a 55-gallon plastic drum, ripped along its waist with a circular saw. He labels each batch with its date of mixture, so that he is always using five-day-old lime putty. Plastic covers the half-barrels.

Bruce's sand is dry — he keeps it under cover — and he brings a quantity of it to the mixing area as needed. He has scales for weighing water, sand, and lime putty and uses them with every batch. Paying careful attention to detail has yielded consistent workable mortar, without cracking — except once, when two-day-old putty was used inadvertently.

5.14: *Kneading the lime putty mortar.*

5.15: *Bruce Kilgore's "production line" mixing area. Note the scales.*

Testing and Consistency

The same *snowball test* described in Chapter 8 works with LPM. If the LPM is too dry, a little water can be added and the batch mixed again. If the mix is too wet, a little dry sand may be added, but not dry lime. Although Bruce and I both prefer a 2.5-to-1 ratio of sand to lime putty, up to a gallon of extra dry sand can be added and still be within the 3-to-1 formula recommended by Mr. Vitruvius 2,000 years ago.

Using Lime Putty Mortar with Cordwood Masonry ...

... is pretty much the same as using the cement-based mixes described in Chapter 8. Normally, final, or *finished,* pointing is done by the end of the workday, although one couple of our acquaintance was very happy with the husband building the cordwood wall one day and his wife coming along the next morning and pointing it, something that is impossible with cement mortars.

Bruce Kilgore's wife, Nancy, did all the building and pointing at Ravenwood, while Bruce and an assistant rushed to keep her supplied with log-ends and mortar. We all agree that the pointing process with LPM is much faster and easier than pointing cement-based mortars. LPM is more plastic and smoother.

Pay Attention to Detail

Lime putty mortar is not for everyone. Success with LPM — and there have been failures — depends upon minimizing variables, as described above. If you are not a detail person, use the more forgiving Portland. Do a test project with LPM before purchasing large amounts of lime and embarking on a 1,500-square-foot house.

Cob As Mortar

Cob is an ancient building material composed of sand, clay, and reinforcing binder, usually straw. Houses with thick solid cob walls have been built in Britain for over 1,000 years, and an entire Essentials book in this series will be devoted to the subject.

Ianto Evans and Linda Smiley have developed and popularized cob building for the modern world through their teachings at Cob Cottage Company in Coquille, Oregon, and through their writings and lectures. Jaki and I were fortunate to have them as guests at Earthwood a few years back, and we were all interested in combining cordwood with cob, something which has come to be called *cobwood.*

Our garage had been recently completed with regular cordwood mortar throughout. The frame was 8-by-8-inch timbers, establishing an 8-inch-thick cordwood wall. We decided to remove a translucent 4-by-7-foot sheet of fiberglass greenhouse covering from one of the panels and replace this "window" with cobwood, figuring we could get by with a little less light in the building, in the interests of science.

We already had good sand on site, both coarse and fine. We found some great Ianto-approved clay at an excavation site nearby, and shoveled it onto my pickup truck. When we got back to Earthwood, Ianto began soaking the clay clumps in five-gallon buckets, to hydrate and soften it so that it would be ready for use the next day.

The missing ingredient was good straw. We had some rotting straw, but Ianto said that wouldn't do, so we ended up using some dry, fairly coarse hay instead of straw. (Straw is preferable, though, because its high cellulose content prevents it from breaking down easily.)

The cob experts could tell by feel that our clay was quite pure, so they recommended a mix that would be about 20 percent clay and 80 percent sand by volume. Some builders may be fortunate in having earth on hand with an already favorable combination of clay and sand. Generally, an earth with clay content of from 10 to 30 percent will yield pretty good cob.

Shovelfuls of our coarse sand and hydrated clay were piled in the middle of a 6-by-8-foot polyvinyl tarp lying on the garage floor. Best to put the sand down first, as the clay tends to stick to the tarp. The ingredients were added at the rate of four parts sand to one part of hydrated clay. After a manageable amount was assembled in the middle of the tarp — about a five-gallon pail full — we turned the ingredients by lifting the edges of the tarp, always folding the goods into the center. This goes better with two people, one on each side of the tarp. Turn until the clay clumps are broken up and the mix has taken on a fairly consistent color.

After turning, the mix is danced on by the cobbers. Jaki and Linda each wore rubber wading shoes, although many cobbers with toughened feet perform this operation barefooted. The purpose of this dance is to drive the tiny clay platelets into the voids between the sand grains. Sand gives the cob its hardness and non-shrink characteristics, while the clay acts as the cement that bonds the material together, giving it strength. The clay can be thought of as natural cement when used in this way.

Water is added to give the cob a good plastic consistency and texture. Then straw (in our test, coarse hay) can be shaken into the mix from the flakes of bales and pressed in with your feet. More straw and water can be added as needed, and you will find it handy to turn the mix over now and again by lifting the corners of the tarp. How much straw? After a while, the cob will feel like a tough cohesive mixture, as opposed to squishy mud. We started out making cob in pretty much the same way as Ianto and Linda would prepare it for a solid cob wall, but soon learned that, for cordwood mud, the straw needed to be chopped to two-inch lengths, or less. We'd put a flake (thin horizontal section) from a bale on the ground, and tilt our rotary lawn mower over it, which produced perfect chopped straw.

We experimented with the *M-I-M method* described in Chapter 8 (except that we were using cob instead of mortar), but we also tested a solid cob joint transversely through the wall. We showed Ianto and Linda how we set the log-ends and found that it was much the same as with mortar — except the cob was stiffer than ordinary cordwood mud. It worked well to use long sausage-like cobs with a cross-section similar to the mortar for which we were substituting.

We had a lot of fun sharing and combining our respective disciplines, and, after two day's work, we had learned quite a bit. Even in the first hours of the test, we were all optimistic that it was going to be a success. Ianto and Linda were happy with the way their cob was performing, and our log-ends didn't seem to mind being laid up with cob instead of mortar. The wall looked like an ordinary cordwood wall, except that the "mortar" was brown rather than gray. The only negative was that we could only build about halfway up the 4-foot high panel. The cob began to slump under additional load. Stiffer cob — on a thicker wall — would be better, as we found in New Mexico in 2015 (see below). But the small, relatively delicate mortar joints of our 8-inch-thick garage panel could not be built with stiffer cob.

Linda showed us how to make a finer cob for pointing. This mix made use of some finer

5.16: *Our cobwood panel in the garage, well protected by overhang, still looks good after 16 years.*

sand that we had on hand, a higher percentage of sand, finely chopped hay, and more water. This plaster-grade cob was easier to point, and the hay strands were easier to hide than when we pointed the regular, somewhat coarser cob. It seemed to work well to recess the cob just slightly more than you want for the finished product, and then, on the same day, apply the finish cob mixture for better pointing. The fine cob, pressed into the cob base under the pressure of the pointing knife, adheres seamlessly.

We have left the panel in place for visitors to see. It is beautiful, with a lovely constellation of brightly colored bottle-ends as a design feature, as well as two large white cedar log-ends. The cob is quite hard, although it can be scratched — barely — with a fingernail.

The panel is 16 years old as I write. It still looks great. There is no deterioration or flaking of materials. And it has a very warm appearance.

In 2015, I conducted a test on our cobwood panel. I sprayed a roughly textured section — there were a couple of hairline cracks, as well — with water, and worked it over with my favorite pointing knife. Instantly, I had a smooth, fresh-looking surface.

Should there be an insulated space between inner and outer cob joints? My gut feeling is that in cold climates the insulated space should be retained. Ianto Evans agrees. Another real plus with the cobwood wall, over solid cob, is that much less cob needs to be mixed. With solid cob joints in a cobwood wall, about 40 percent as much cob needs to be mixed as with an all-cob wall. With the insulation in there, the amount of cob needed would be more like 25 percent. The wooden portion of the wall has a higher R-value than does solid cob of the same thickness.

5.17: *The cobber's dance. Experienced helpers made our cob for us at the 2015 Natural Building Colloquium.*

Our Latest Cobwood Wall Building

Jaki and I were invited to demonstrate "cobwood" at the 2015 Natural Building Colloquium in Kingston, New Mexico. As we were keen to do further experimentation and there were young cobbers willing to assist, we were happy to oblige. The project was to build an exterior wall 18 inches wide and about 2 feet high. Although such a wall does not need insulation, we decided to demonstrate how this could be done for a home. We made a M-I-M stick demarcated into 6-inch sections for mortar, insulation, and mortar. The mortar, of course, was cob, so I suppose we should have made a C-I-C stick.

There was little available sawdust for the insulation, but the cob people had a wonderful machine on site that chopped straw into one-inch pieces, perfect to insulate the gap between the two cob joints — and also perfect as the reinforcing binder in the cob itself. With experienced volunteer help, we didn't even have to make the cob. Our assistants had us check the consistency of their mix, so that we had the right "slump" or stiffness to lay up the wall. The mix was about 25 percent clay and 75 percent sand, with the straw for extra tensile strength — like the glass fibers put into reinforced concrete. With all-cob construction (no cordwood), the straw is used at much longer lengths. The beauty of the short lengths is that pointing the surface is very much easier.

With cob mortar always at the ready, work progressed smoothly. We would place our two 6-inch-wide by — roughly — 1.5-inch-thick cob joints, install our straw insulation, and then start placing log-ends in the usual way, careful to get the wall into the random rubble style by choosing pieces of varying diameters.

5.18: *Cobber Jaki having fun getting her hands muddy.* Credit: Rob West.

In New Mexico, we learned that the chopped straw machine is great and would enable a cobwood builder to produce any amount of cheap insulation if they had straw on hand, which they would have to in order to make the cob. The chopped straw, it seemed, would be a very effective insulation, and it was easy to install in the wide space in the middle of the wall. Finally, we were impressed with the mass and power of the wall. I think such a wall could be load-bearing in a non-seismic area, but the builders would still have the downside of working out in the open. Care would need to be exercised to protect the wall/building from rain. We only had time to build up about 18 inches in a day, but I am confident that we could have built twice that high with no slumping of the cob, which we had experienced years earlier on our relatively delicate garage wall.

5.19: *Our cobwood wall demo in New Mexico.*

Cob As Mortar: Summarization

1. Cobwood is appropriate for builders with access to good clay. We once experienced a failure at a workshop where our hosts provided what they thought was good clay. Indeed, the cob/mortar we made seemed to perform very well in the building process. But it never set up. In fact, it turned crumbly in time. Fortunately, we had also taught regular mortar at the same workshop, so not all was lost. Be sure of your clay. If in doubt, do a test panel and observe it after two weeks.
2. If you've got access to good clay, cob is kinder to the planet than cement-based mortar.
3. Thick cobwood walls of 16 inches or better *might* support a considerable load in

non-seismic areas. This seems to be the case in South Korea, where cob and cordwood have been combined in round houses, but without an insulated space.

Papercrete, or Paper-enhanced Mortar

Chapter 12 of *Cordwood Building* (New Society, 2016) describes in some detail three cordwood builders who used variations of *paper-enhanced mortar,* or *PEM* for short: Jim Juczak, Alan Stankevitz, and Tom Huber. Their techniques and recipes were a little different, but all were successful. Here are synopses of their methodologies:

Jim Juczak's mortar is "made of paper sludge — 80 percent by volume — that I get (free) from a local paper mill. The other 20 percent is Type N masonry cement." Jim made his PEM in five-gallon buckets, 4 parts wet paper sludge to 1 part cement, and mixed it right in the bucket — in about a minute! — with a "heavy duty spackle blade on a half-inch drill." There is no sand in Jim's mix, so it is very light in weight when it is cured, which takes a month or so for all the moisture to transpire out of the wet paper sludge. In fact, this papercrete mortar is so good as insulation that he simple pours it out onto the wall in progress without benefit of a special insulation cavity — no M-I-M-stick. Jim cautions not to lay up cordwood masonry more than two feet high in a single day. "The PEM is quite Jello-like, and the wall will start to lean in various directions if you build too high, too fast."

Alan Stankevitz, who coined the term *paper-enhanced mortar,* uses sand in his mix. After much experimentation, the mix he likes best is, in parts by volume: 2 parts drained slurried paper, 2 parts fine sand, 1 part Type N masonry cement. While not as light and airy as Jim's mortar, Alan's is harder and stronger. As his house is of double-wall construction, with foam insulation between two 8-inch cordwood walls, the lesser insulation characteristic of his mortar is not an issue. In fact, the extra thermal mass on both sides of the insulation is probably a big plus in keeping internal temperatures stable.

Tom Huber buys GreenFiber cellulose in compressed 2.2 cubic foot bales, which is then fully soaked in water. He, too, experimented for "several years" and has been very pleased with his current mix, which is: 5 gallons moistened cellulose, 2 gallons Type S masonry cement, 2.5 gallons mortar sand (also called *mason's sand*), and 2 gallons hydrated builder's lime. Tom shares his mixing details: "When ready to make up a wheelbarrow batch, I first drain off and squeeze out most of the water using a small screen placed over one of the barrels and fill a five-gallon pail with the moistened cellulose. I then dry-mix the sand, cement, and lime (from measured pails), and then knead in the 5 gallons of cellulose by hand. It is important that the same proportions be used for every wheelbarrow load, in order to have a uniform mortar color when it dries."

Tom says the biggest drawback to his mortar is: "It stinks! It literally smells from an ammonia type off-gassing that occurs between the fire retardant in the cellulose and the chemical composition of the hydrated lime. Common fire retardants added to cellulose insulation include borax, boric acid, and ammonium sulfate. After I first smelled ammonia in a PEM mix, I Google-searched and found similar reports among papercrete users. The presence of ammonia produces little risk when mortar mixing in an open-air environment, but a protective ventilator mask should be used in tight interior spaces. Once the mortar sets, it no longer smells unpleasant in any way."

Hempcrete ...

... is beginning to gain popularity, particularly in Canada, untroubled by archaic substance laws. Chris Magwood has devoted an entire book in this series to hempcrete, *Essential Hempcrete Construction* (New Society, 2016). I will not even attempt to distill his comprehensive work down to a paragraph or two, especially as I have never used the material, nor seen the results in person. However, in the color section of his book, in the caption of a picture of a cordwood masonry interior wall, Chris says: "The fit of cordwood and hempcrete is excellent, and could be used for exterior walls as well."

Part 4: Insulation

Some Options

A cordwood wall derives its exceptional thermal characteristics from the insulated space between the inner and outer mortar joints. If this space is not insulated, the house will be expensive to heat when it is cold outside. Staying cool in hot weather depends on good insulation, as well.

There are several choices for insulation in this space. Jaki and I did our first three buildings with fiberglass from a roll, but it was nasty stuff to work with. If you use it, wear eye protection and a respiratory mask. Also, it has a high embodied energy in its manufacture, and, if it mats down with moisture from any source, it may or may not fluff back again. Vermiculite, perlite, and other loose-fill insulations can work quite well, but can be costly. Shredded expanded polystyrene (bead board) may seem, at first, to be a good way to recycle materials, but, thanks to static cling, it is very difficult to direct the stuff into the insulation cavity. The little beads go everywhere except where you want them, and the slightest wind is a disaster.

Sawdust

Around 1980, Jaki and I switched over to using treated sawdust for the insulation cavity and have been very pleased with the results ever since. Sawdust is cheap, makes use of a waste material, and has an R-value of about R-3 per inch, about the same as fiberglass. And it is easy to pour into the cavity with a small spouted bucket, or, in tight spaces, with a tin can.

The best sawdust to use is from the same species of wood that are good for log-ends, but even hardwood sawdust will work, although perhaps at a somewhat lower R-value. Use sawdust from a sawmill that uses a circular saw — it pours more easily than fine sawdust from sanding, such as from a cabinetmaker's shop. Wood shavings are not as good as insulation, or as easy to use, but, if sawdust is unavailable, they might be a viable choice with cavities 8 inches wide or more.

Sawdust is nontoxic, has low embodied energy and makes use of a waste product.

Making the sawdust insulation is covered in Chapter 8. Some other green insulation options are covered later in this chapter.

Part 5: Other Insulation Options

In the past few years, materials other than sawdust have been used with cordwood for insulation. Ivan McBeth, building in Vermont, used cellulose insulation which comes in a 30-pound bag. His cavities were about eight inches wide, so it went in quite easily. Ivan told me that his lady friend reported some allergic rash when using it, although he did not, so rubber gloves might be advised in the handling. The price per R is favorable on cellulose insulations compared with other manufactured products. The insulation is "green" in that it uses recycled paper and has a lower embodied energy in its manufacture than fiberglass (0.94-3.3 MJ/kg for cellulose versus 28 MJ/kg for fiberglass according to the Inventory of Carbon & Energy [ICE] Database). Treatment with borates protects against fire and insect infestation.

Mortar aggregate vermiculite, which comes in bags, has been used by cordwood builders in Canada instead of sawdust. Be sure that your vermiculite comes from an asbestos-free mine.

Perlite, another loose-fill insulation, will work well, too, but these products will add considerably to the house cost when compared with lime-treated sawdust.

With *cobwood,* discussed above, we used straw which had been chopped into one-inch pieces. Others, particularly in Canada, have used hemp insulation for the purpose. Both have worked well, and I would think yield R-values similar to sawdust, right around R-3 per inch, give or take.

Spray Foam

Three of the most beautiful and energy-efficient cordwood homes ever built have sprayed-in foam as insulation, both as the fill in a double-wall system, as well as sprayed into the insulated mortar joint during construction of a regular single-wide cordwood wall. The owners are very pleased with their results, but it must be recognized that the use of spray foam, even the partially "soy-based" foams, have environmental drawbacks. While I will not be discussing the use of spray foam in this book, those who want to learn about it can go to Chapter 7 of my *Cordwood Building: A Comprehensive Guide to the State of the Art* (New Society, 2016) for detailed information.

CHAPTER 6

Design Options for Cordwood

THERE ARE THREE BASIC WAYS that cordwood masonry can be used in *single-wall* construction, defined as the same log-end exposed on both the exterior and the interior of the building. They are: (1) cordwood as infilling within a strong timber frame, (2) load-bearing curved-wall buildings, and (3) load-bearing cordwood walls built between *stackwall* — or *built-up* — corners, all discussed below.

In addition, several builders have used *double-wall* construction, in which interior and exterior cordwood masonry are separated by a continuous insulated space. (The Raddush Fort in Chapter 1 is an example of a double-wall building from nearly 1,000 years ago.) Typically, double-wall cordwood homes consist of two 8-inch-thick cordwood walls separated by 8 to 10 inches of insulation, be it sprayed foam, fiberglass, or something else. Usually, double-walled cordwood homes are rectilinear in shape, and are built under the umbrella of a timber frame with its roof on. Bruce and Nancy Kilgore's Ravenwood home in Saranac, New York, is an exception, being triangular; it is featured in its own chapter of *Cordwood Building.*

Foundations

Cordwood masonry can be built on conventional foundations, but the footings should be at least as wide as the cordwood wall itself; a bit wider is better.

Foundations can be expensive, particularly if you hire them out. A good foundation for cordwood masonry is Frank Lloyd Wright's *floating slab*: concrete built on a pad of good percolating material such as coarse sand, gravel, or crushed stone. A floating slab can be built by an owner-builder (backed by some basic research and three or four fit helpers). Beware, though, of attempting to build crawl spaces; they are, literally, a pain in the neck.

With basements, you need to plan ahead so you can put electric and plumbing under the slab. But think hard before deciding on even having a basement. Basements are expensive and yield low-quality space. We discovered this for ourselves with our Log End Cottage. The basement was half the cost of the building — but it got only five percent of the use. If you must have a basement, go the extra mile on good waterproofing, exterior insulation, light, and ventilation. Aim to create a high-quality earth-sheltered space, such as the lower story at Earthwood. For more information, you might want to read my book *Earth-Sheltered Houses* (New Society, 2006).

For my friends who — rightly — see concrete as a carbon-producing behemoth, there is the option of putting a stone masonry footing around the perimeter, and installing a wooden floor. Stone masonry is beautiful, satisfying, and makes use of an indigenous material, but it is much more difficult and time-consuming than cordwood masonry. Been there, done that.

Or, cut concrete in half by doing just the footings in poured concrete — a *floating ring beam* — and, again, have a wooden floor over a vented space. Exhaust, plumbing, and electric can run through this space, but, if it is ever necessary to access it, you have to do it by removing boards, rather than going through a crawl space. With either the floating slab or ring beam, there is the option of installing a single course of solid (non-cored) concrete blocks mortared on top of the foundation. This keeps the first course of cordwood an extra few inches off of the ground — advisable in very wet climates — which is what we did when we built with a client on Hawai'i, the Big Island.

Single-wall Cordwood Methods

1. Cordwood as infilling within a timber frame

This is the most common way that cordwood masonry is used nowadays — for two very good and valuable reasons.

One, getting a timber frame up first enables the builder to cover the building site with the umbrella protection of the roof. Cordwood masonry benefits from good overhang protection anyway — I recommend 16 inches all around — and this overhang gives further protection during the labor-intensive — dare I say slow? — cordwood-building process. It is an easy matter to stretch a tarp out from the roof to gain another six feet of protection against rain. With the other two methods — described below — it is necessary to build a temporary structure to build under, or, at the least, a movable protective framework, as we did at the round Earthwood house.

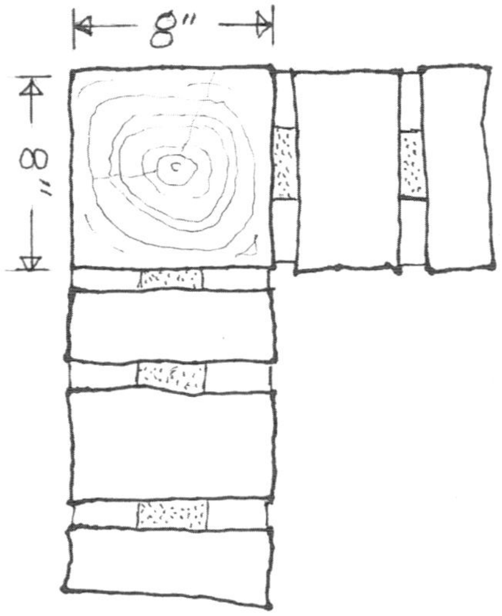

6.1: *Single-wide 8-inch-thick cordwood walls are laid up against the side edges of an 8×8 corner post.*

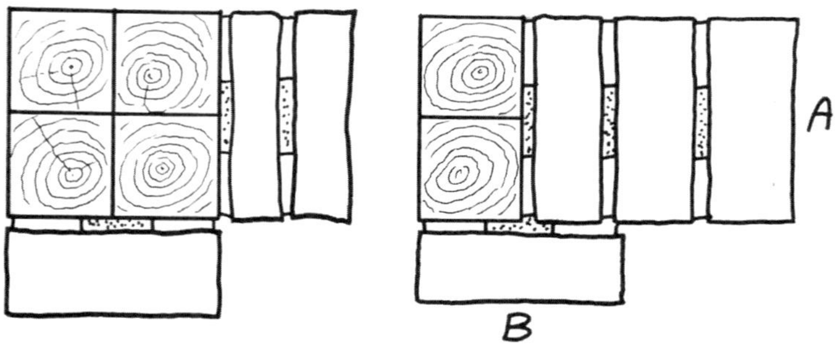

6.2: *Instead of four 8×8 posts in the corner of a 16-inch wall (*left*), use two. Then build wall A before wall B (*right*).*

Two, the timber frame is a positive selling point for builders who need to get the local code enforcement officer on board. The CEO will understand the structural advantages of the timber frame. He or she may have to be convinced about the load-bearing capacity of an unsupported cordwood wall. Despite the Raddush Fort, which lasted for hundreds of years; existing stackwall-cornered barns in Wisconsin dating back to the 1800s; and our two-story round Earthwood house supporting a living roof load of 185 pounds per square foot since 1981, we have no authoritative compression tests on a load-bearing cordwood wall.

My wife and I love round buildings, for reasons to be described, but if we were to build Earthwood again, our only major change would be to build it as a *16-sided* timber frame — a *hexadecagon* — with its 16-sided umbrella roof to protect us during cordwood work. A hexagon and an octagon look like what they are, but a hexadecagon looks very much round, and has the same spacial advantages of round, while giving 8-foot straight walls that are more accommodating for furniture, bookshelves, and countertops.

Granted, timber framing is an additional process to the cordwood, but it will save time and trouble in the long run. While I have the highest regard for traditional timber framers, using time-honored jointing methods, I lack the skill, patience, money, and special tools required to do it myself. Instead, I practice what I describe in *Timber Framing for the Rest of Us* (New Society, 2004), whereby the timber frame is erected using commonly available — and relatively inexpensive — mechanical fasteners. In fact, this is the way most farmers, owner-builders, and even contractors actually do it. The net effect is that you see the timber framing — commonly made from rough-cut 6-by-6 inch or 8-by-8 inch timbers — either on the inside or the outside, your choice. It is not necessary, incidentally, to build a 16-inch wide timber frame for 16-inch-thick cordwood walls, as can be seen in the illustrations. The actual joining of timbers is beyond

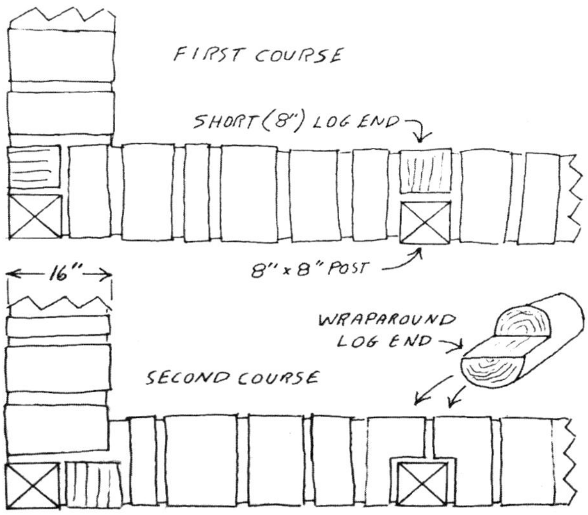

6.3: *A single 8-by-8 inch post will suffice in the corner or the side of a 16-inch cordwood wall, by weaving successive courses as shown. A wraparound log-end is made like the shelf unit seen in Figure 9.6.*

Impact of Perimeter Shape on Area

Shape A: the circle. The choice of other building species. The most space per foot of perimeter.

Shape B: the square. The most efficient rectilinear shape, seldom seen today.

Shape C: the rectangle. The most common house shape today. Why?

Shape D: the 1950s mobile home. The longer and narrower we make it, the less space we enclose. We could build 59 feet long and 1 foot wide and have 59 square feet.

Shape E: the architect gets involved. If those two "inner" walls had been left on the outside where they belong, we'd have the efficient Shape B. The roof is more complicated to build with Shape E and 225 square feet are lost.

Shape F: the "hockey rink." Many people are building cordwood homes like this, and not just in Canada. Still over ten percent more space than the most efficient rectilinear shape (B), which almost no one builds. The roof is not complicated if a radial rafter system is employed for the half-circles. The radial rafter corresponding to the internal arrows on the diagram is also the first of any number of parallel rafters for the rectilinear section.

In fairness, it must be pointed out that if you enclose more space, you will spend more time and money on roofing, foundation, and flooring. However, these components go faster than the labor-intensive cordwood walls. The important point here is that, with a circle, you can get any desired floor area by building less perimeter wall. Also, with less skin area, the home is more efficient to heat. Just thought I'd share this with you.

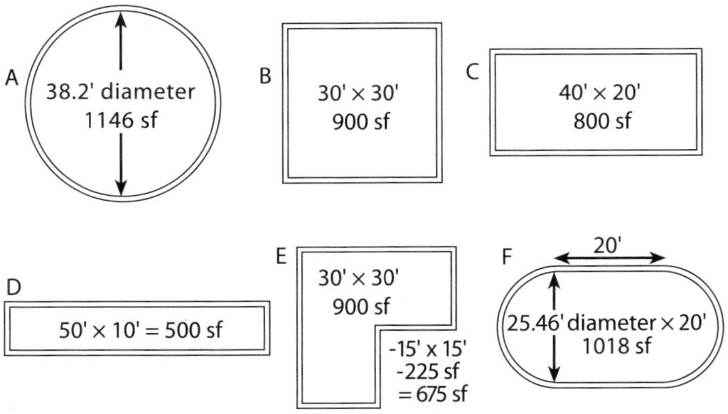

6.4: *All six of these house shapes have an interior perimeter of 120 linear feet. Look at the varying square footage figures.*

the scope of this book, and the reader will find the information in either the book just mentioned, or, for the purist, one of the fine books on traditional timber framing by craftsmen like Will Beemer, Steve Chappell, and others.

6.5: *We built the first story of our 22-foot diameter Mushwood Cottage under a 29-foot diameter geodesic dome, then re-erected the dome at the second level and lived in the mushroom-shaped cottage for 25 years.*

6.6: *When the dome began to deteriorate, we built a second story of cordwood masonry with a conventional roof. It can be seen on the cover of my book,* Cordwood Building *(New Society Publishers, 2016).*

2. Load-bearing curved single wall

Usually, a curved-wall building is truly round, like our Earthwood home, as well as our sauna, office, and bookstore. Our summer camp, Mushwood, is also truly round, with the cordwood load-bearing. There have been other curved-wall shapes employed, such as the spiral and the hockey-rink shape, and not just in Canada. But it is safe to say that more than 90 percent of curved-wall buildings are round.

A compelling advantage of a truly round building is that round is the shape which encloses the most space per unit of perimeter. The amount of wall materials required, as well as the cost and the hours of actual cordwood masonry construction are also less than with any of the rectilinear shapes. See Figure 6.4.

Not everyone likes a round building, although visitors smile when they see one, inside or out. Some fear that a round house will be more difficult to build, even though a fair amount of so-called "primitive" houses, ancient and present day, in Asia, Africa, and Mayan America are built that way. Animal building species — birds, bees, beavers, and others — also choose round for ease and efficiency of building as well as the thermal advantage of having the least skin area to enclose desired space. Yes, modern buildings rely heavily on sheet goods, such as plywood, plasterboard and the like, but these are not used in the cordwood wall. Various methods of keeping the correct curve — and the verticality — of the cordwood walls are covered in Chapter 8.

A downside of load-bearing cordwood walls, curved wall or otherwise, is that the work might be exposed to the elements during construction. We built our 20-foot diameter office building — and both stories of our Mushwood summer cottage — under a geodesic dome. See Figures 6.5 and 6.6.

Our 10-foot diameter sauna was built under a large tarpaulin stretched out from the

Earthwood house. But at Earthwood itself, we had to keep moving a temporary protective shelter around the circle as we built; we carefully covered all work with weighted plastic to stop rain from getting into the insulation space.

3. Load-bearing stackwall corners

In this method, used with buildings having right-angles (square, rectilinear, L-shaped, etc.), the corners are built first, with wooden blocks called *quoins* (although at least one builder has used decorative pre-cast concrete block corners). This method was popular with 19th-century barn builders, as well as Canadians who saw it as a good way to build cordwood walls two feet thick or more for their cold climate. Fewer stackwall-cornered homes are built nowadays, although some lovely examples still pop up now and again.

The quoins are most commonly made of sawn squared timbers, such as 4×4s, 4×8s, 6×6s and the like. Two quoins are mortared up side by side, or with a log-end spaced between them, to give the desired width of wall. At Earthwood, for example, we have some short stackwall corners made of 4×8s, side by side, establishing the 16-inch cordwood wall thickness. One of the drawbacks of the stackwall corner method is the use of far more regular sawn lumber than would be required for a single squared post. This disadvantage has been overcome by some builders, including 19th-century barn builders, who have used quartered logs of a regular size for their quoins.

Do not be tempted to use round logs as quoins. Corners made from rounds are very much less stable than using quoins with at least one flat side (although two opposite flat sides are best).

There was not a lot of change in stackwall corners for many decades. Then two late 20th-century developments appeared at about

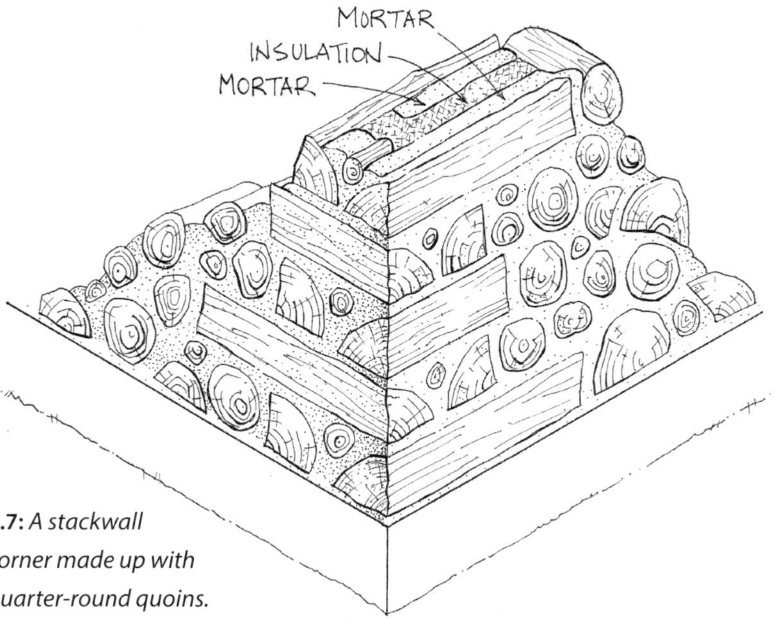

6.7: *A stackwall corner made up with quarter-round quoins.*

6.8: *Two Lomax corner units, made from 4-inch-square timber stock, are stacked one upon the other. Note the tie pieces and chainsaw grooves for better friction bond to the mortar, not shown in this model. The short block is simply a decorative filler piece and helps retain the sawdust insulation which will be placed in the continuous inner cavity of the corner. In the background is a stackwall corner at Earthwood, made from 4-inch by 8-inch quoins for this 16-inch cordwood wall. The wall continues above with the cordwood supported by a "double-wide" pair of adjacent 8-inch-square posts.*

the same time; each improved on the technique. Gary Lomax of New Brunswick designed and used pre-built corner units which came to be known as Lomax Corners. These corner units are composed of two regular squared quoins tied together with 1-inch by 1-inch or 1-inch by 2-inch pieces, seen in the Figure 6.8. The advantages of the Lomax system are: (1) stronger than individual quoins not mechanically tied together; (2) faster, easier, and more accurate laying up and leveling the corners; and (3) each corner goes up at the same rate, 6 inches in the example shown in the picture. Sixteen Lomax units, each rising 6 inches, totals 96 inches, or exactly 8 feet.

The other development involved using two different lengths of quoin, whether or not Lomax units are employed. For example, half of the quoins can be 24 inches, and half can be 30 inches. As a corner is built, use two consecutive (adjacent) courses of 30-inch quoins or units, then two consecutive units of 24-inch units. This technique helps prevent a vertical shear crack from developing in the corner, where the pressure on the foundation is greater than along the sidewalls.

Upsides to stackwall corners: (1) This technique is an option if you haven't got access to affordable heavy timbers for a timber frame. Quoins can be made from quartered logs or by ripping two slabs off a log with a chainsaw. (2) Any thickness of wall can be built. (3) Done with care, stackwall corners can be quite beautiful.

Downsides to stackwall corners: (1) It takes considerably more milled lumber to build good stackwall corners than to make a single 6×6 or

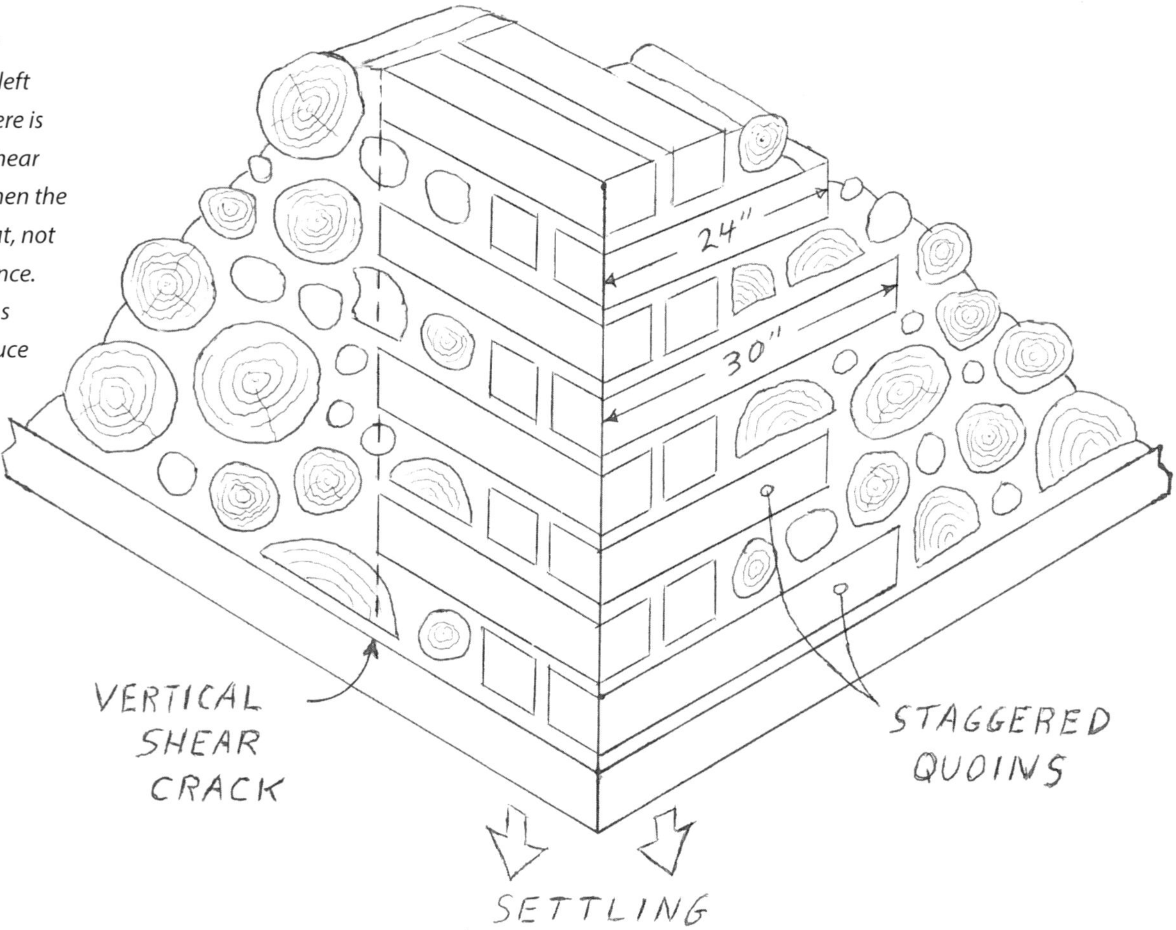

6.9: *If all quoins are the same length, as on the left side of this drawing, there is a danger of a vertical shear crack forming if and when the corner settles somewhat, not an uncommon occurrence. On the right side, quoins of different lengths reduce the chances of a direct perforation.*

6.10: This small pump-house in North Carolina was made with single-wide quoins, cut from tulip poplar with a chainsaw, then water-sealed to minimize water absorption from mortar to wood. The staggered length of the quoins is clear in this image.

8×8 post. In common with a timber frame, the tops of the completed stackwall corners need to be tied together with a strong plate (girt) system, upon which floor joists, trusses, or roof rafters rest. (2) Stackwall corners are more time- and labor-intensive. Some of this is the constant need to cover and uncover the work in inclement weather. (3) The roof cannot go on until all the cordwood is completed. All building components, including electric, are exposed to the elements. (4) Sawn quoins are much more absorbent of moisture from mortar (or rain) than regular cordwood masonry, so swelling and uplifting of the corners can result. This danger is greatly reduced by the application of a good waterseal on any sawn quoins prior to laying them up.

No matter what quoin material is chosen, the basic order of events with stackwall corners is the same. Install and plumb your door jambs (frames) first. Then, with a simple four-cornered

6.11: A beautiful stackwall cornered addition in Peru, New York, built without staggered-length quoins.

building, lay up all four stackwall corners — mortar, insulation, quoins — to a comfortable stable height, say between 30 and 48 inches. This height is called a *lift*. Take care that the corners are going up level and plumb on both sides of the corner. Then, stretch a mason's line from corner to corner. The clip which holds the line keeps it one-half inch from the quoins. Now build the regular cordwood masonry between corners, always keeping log-ends a half-inch back from the tightly stretched line. You can see a mason's line in Chapter 8, Figure 8.16 and 8.17.

After the entire building is up to the height of the first lift, build all the stackwall corners up to the height of the second lift. This could be 60 inches, or even full wall height. Continue with the cordwood masonry between corners. Window bucks (frames) are usually installed during the second lift (even earlier in the case of tall windows).

A typical 90-inch-high cordwood wall can be done in two or three lifts. Then the corners need to be tied together with a double-thick wooden plate system, as per Figure 6.12.

With all three single-wall techniques, the actual building of the cordwood masonry walls is the same, and is described in detail in Chapter 8.

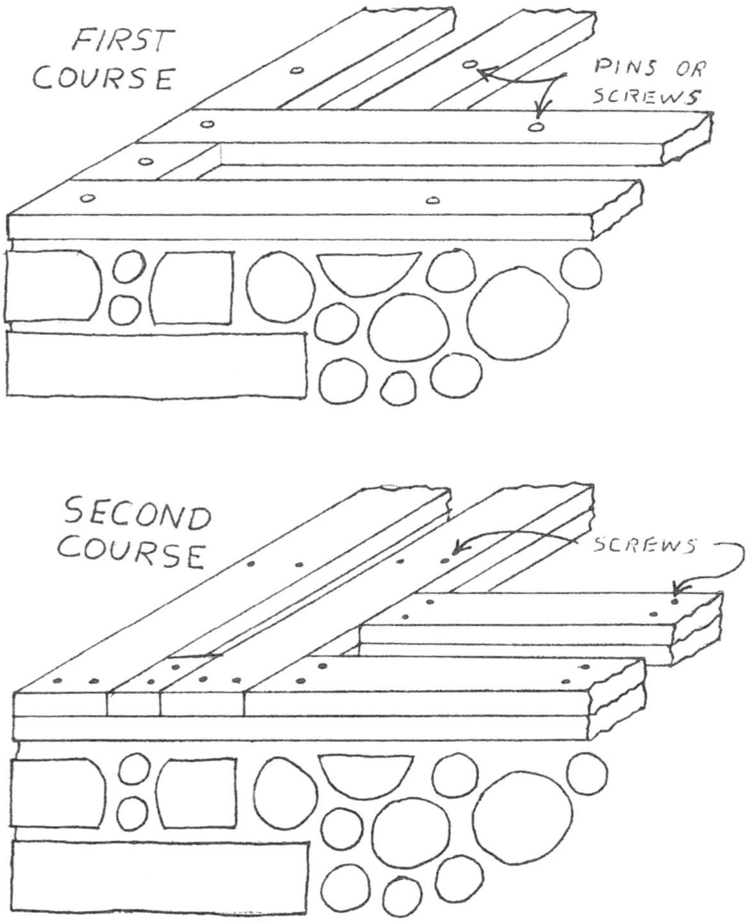

6.12: *A double-thick wooden plate system, with overlapping courses, ties the stackwall corners together.*

The Double-wall Technique

Popularized by Saskatchewan native Cliff Shockey, the double-wall technique involves the building of two separate cordwood walls all around the building: an inner wall and an outer wall; these are separated by a heavily insulated space. Typically, both cordwood walls are 8 inches thick, and the insulated space is also 8 inches, yielding a well-insulated 24-inch-thick wall, with plenty of useful thermal mass on each side of the insulation. Cliff built his first home in 1977 and his second — larger — home a few years later using stackwall corners and cordwood for the outer wall, and cordwood infilling between 8×8 posts on the inner wall. Cliff says: "In 1985, I decided on post-and-beam framing for both the inner and the outer walls of the insurance office, and did the same thing in a 392-square-foot (36-square-meter) addition to our second house in 1990. [This method] is faster and easier and enables you to get the roof on first and then work under cover."

Cliff used a solid mortar joint — having no insulated space — of three parts sand and one part masonry cement for both cordwood walls, as have most other double-wall builders. Bruce and Nancy Kilgore, though, used the regular M-I-M cordwood method

(Mortar-Insulation-Mortar) at Ravenwood, thus saving on mortar mixing and even further improving the insulation value of their 26-inch-thick wall.

Cliff says: "I like to put 5⁄16-inch particle board (or any inexpensive sheathing) on the inside of the exterior wall. This acts as backing for the (fiberglass) insulation batts to come. It also serves as a barrier to help keep bugs and mice out of the insulation cavity." Cliff used 8-inch fiberglass batts "because they are fairly rigid. I find that they will stand on end without sagging down the wall. Next comes the vapor barrier ... important for making an airtight, draft-free home. The vapor barrier is fastened to a 2×8 pressure-treated plate at the base of the wall, and also to the inner post-and-beam frame, the top plate, and all window and door frames. All seams in the vapor barrier must be sealed together with ... sealant over solid backing. Make sure that every seam is sealed before the walls and ceilings are finished, as it is impossible to get to it later. ... One advantage of the double-wall system is that any irregularities in log-end length or straightness of cut can be hidden out of sight toward the center of the wall."

The double-wall technique "isn't twice as much work as regular cordwood masonry," says Cliff (even though two walls are being built instead of one, only two surfaces are pointed, not four, saving time). Plus, Cliff was a very efficient builder and his pointing method was "a rough pointing with rubber gloves. Later, we cleaned the log-ends of loose mortar with an electric rotating wire brush and then sprayed the wall with a spirit-diluted mixture of polyurethane. This brings the color out in a very attractive way, and provides a surface that is easier to clean."

Two of the most beautiful — and energy-efficient — double-wall cordwood homes I know of are Bruce and Nancy's Ravenwood (already mentioned) and a two-story hexadecagon built

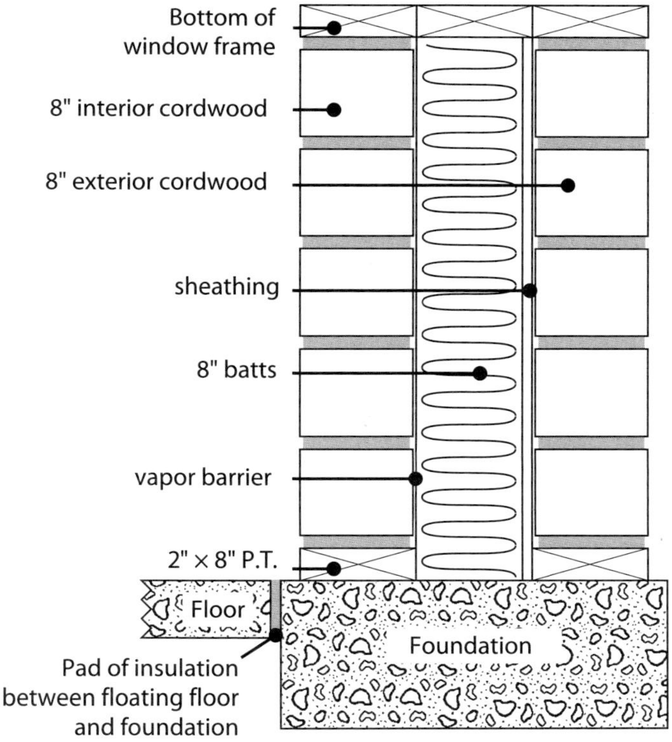

6.13: *Cross-section of Cliff Shockey's double-wall technique.*
CREDIT: ROB PICHELMAN.

6.14: *Plan view of simple post and beam frame for the double-wall technique.*

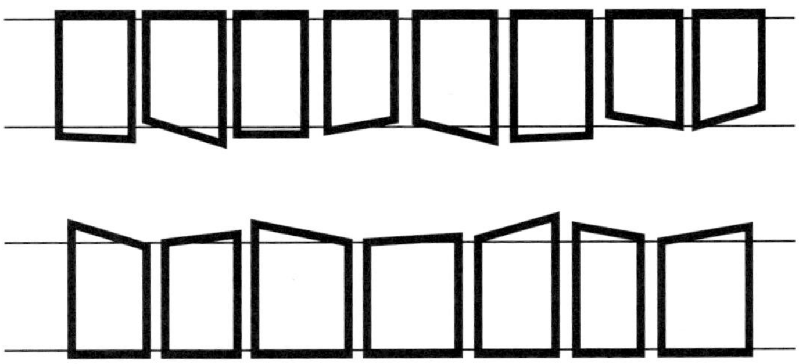

6.15: *Plan view of double wall. Log-ends do not have to be cut perfectly in order to keep the interior and exterior surfaces straight.*

by Alan Stankevitz near La Crescent, Minnesota. Alan's house took him eight years to complete, and the cordwooding at Ravenwood took five years. In fairness, both parties had comfortable places to live and regular jobs while they built, so it was part-time cordwooding. The reader must decide whether the extra time of double-wall building will be worth the effort.

Door Frames

With all styles of cordwood masonry, it is necessary to get your door frames in place before the cordwood can commence. Cordwood masonry is heavy and plastic during its first couple of days of curing — even longer with lime putty mortar and cob — and it places a considerable lateral (sideways) load on the doorframes, so they need to be made of substantial material. The frame consists of two side members, called *jambs,* and a header tying them together on top. My personal preference, which I've used on at least a dozen buildings, is full-sized 4×8 (4 inches by 8 inches) timbers. Architecturally, and structurally, they are in keeping with a strong timber frame, but they also work well with stackwall or curved-wall construction. (Warning: Even a full-sized 2-inch-thick frame may not be strong enough, and a "2-by" bought from a box store or most lumber yards is actually only 1.5 inches thick — and almost certain to bow in and make it impossible to use the door.) For 8-inch cordwood walls, use the 4×8s as is. For 16-inch walls, double them up, so that the door frame is actually 4 inches by 16 inches. For a 24-inch-thick single (or double) wall, use three side by side. With other thicknesses, you may have to choose different dimension timbers, such as 4×6, 4×10, or 4×12.

6.16: *Top of door frame at Earthwood, made from doubled 4-inch by 8-inch jambs and header.*

Earthwood, for example, has door jambs composed of two 4×8 timbers scabbed together with a vertical 1-inch by 6-inch key piece along its full height. The key does double duty; it ties the inner and outer door jambs together for construction, and it provides a positive locking key for the cordwood masonry later on (this is discussed in Chapter 8). The door's header should be made of the same materials.

I always like to extend my header out about four inches proud of the door jambs on each side, as seen in Figure 6.16. It looks good, yes, but it also lends a little extra bending strength to the header, particularly on wide doorframes, such as a six-foot sliding glass door unit. In fact, on wide doors, I will extend the header as much as eight inches both sides. Fasten the header to the posts with two strong structural screws, such as TimberLOK or — my favorite — GRK Fasteners.

The various methods of fastening posts and heavy door jambs to a concrete foundation are beyond the scope of this book, but are discussed in detail in Chapter 4 of my *Timber Framing for the Rest of Us* (New Society, 2004, still in print). But if you are somewhat confident with your timber framing skills, Figure 6.17 might be useful.

When building cordwood within a strong timber frame, it is worthwhile to plan your post

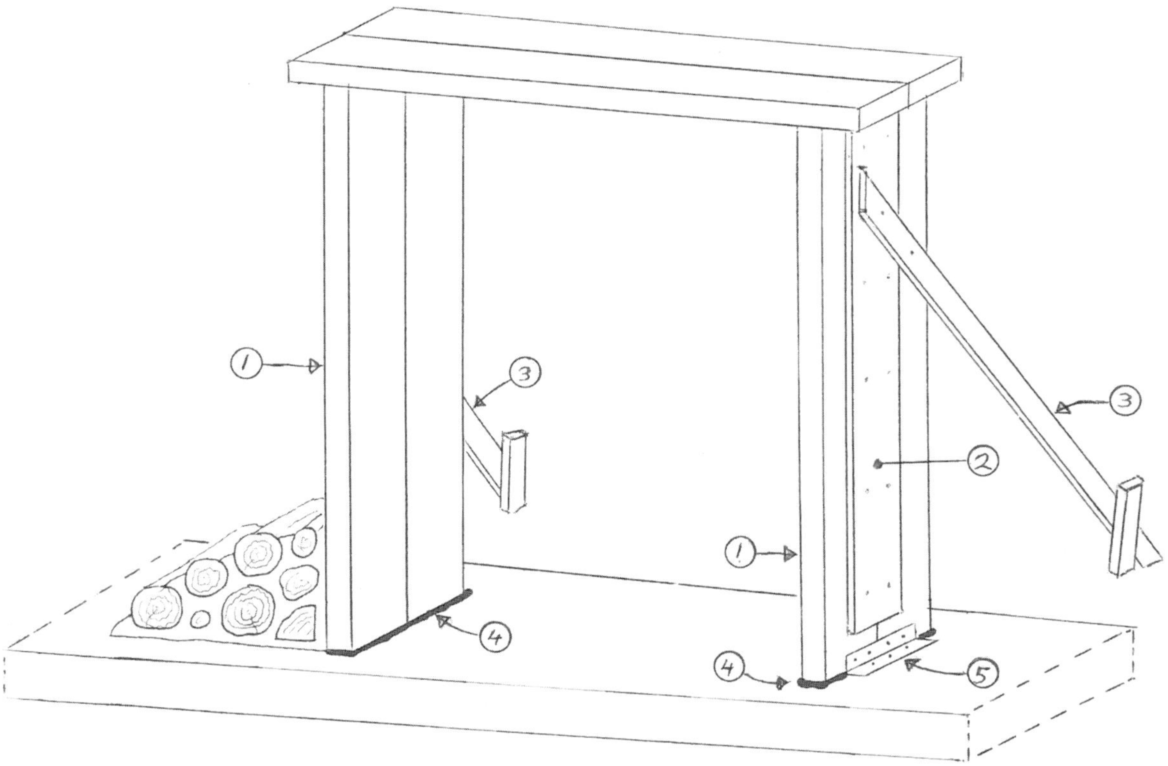

6.17: *A frame for a 6-foot-wide sliding glass door unit.* **1)** *Double-wide 4×8 timbers for a 16-inch cordwood wall.* **2)** *1×6 boards tie the 4×8s together and act as key pieces to lock the frame into the cordwood wall.* **3)** *Temporary support braces plumb and steady the doorframe.* **4)** *A damp-proof course between door jambs and the foundation can be a Bituthene waterproofing membrane, asphalt roll roofing, or scrap 240-pound asphalt shingles.* **5)** *2-by-2-by-¼-inch angle iron fastened to the footing with lag screws fastened into lead expansion shields and to the frame with lag screws.*

spacings to accommodate door frames and windows, as we did at Log End Cottage and Log End Cave, as well as several of the outbuildings at Earthwood. At the Cottage, our eight-by-eight-inch posts served double duty as door jambs. The heavy girt that joins the top of the sidewall posts can also serve as the door header (or the tops of window frames). A lot of time and money is thus saved at the design stage.

6.18: *Log End Cottage, built in 1975. Doors and windows are enclosed by the timber frame itself, mostly 8-inch by 8-inch recycled barn beams.*

Four examples from four different shapes of buildings, are shown in Figures 6.18 to 6.21.

Although windows can be framed by the timber frame, as in Figures 6.18, 6.19, and 6.21, heavy frames — called "window bucks" — can also be floated in a cordwood wall, like the diamond-shaped ones in Figure 6.20. The technique is explained in Chapter 8.

6.19: *Strawbale (and cordwood) guest house at Earthwood. The door is exactly framed by the 4-inch by 8-inch jambs which extend from the foundation to the 8-inch by 8-inch girt.*

6.20: *Our round office building has walls 12 inches thick. The door frame is made from full-sized 6-inch by 12-inch timbers. Note that the header carries a few inches into the cordwood masonry. Yes, this door opens out.*

6.21: *Stoneview is an octagonal guesthouse with eight 8-inch by 8-inch girts joined over eight special posts. Both the door jambs and the long window frames (all 4-inch by 8-inch stock) extend from the floating slab to the underside of the girts. Full framing details are in my book* Stoneview: How to Build an Eco-friendly Little Guesthouse *(New Society Publishers, 2008).*

Electric Wiring with Cordwood

At the design stage of your cordwood building, consult your local electrical code and electrical inspector to determine whether there are any complexities or special concerns that need to be addressed.

Wiring for wall outlet circuits may be laid in the insulation cavity during wall construction. Flexible wall conduits are recommended for this (see Figure 6.22). At least one cordwood builder, however (Ed McAllen of Galesville, Wisconsin), used direct-burial Romex conductors in the center of his 16-inch-thick cordwood walls and met with code approval because

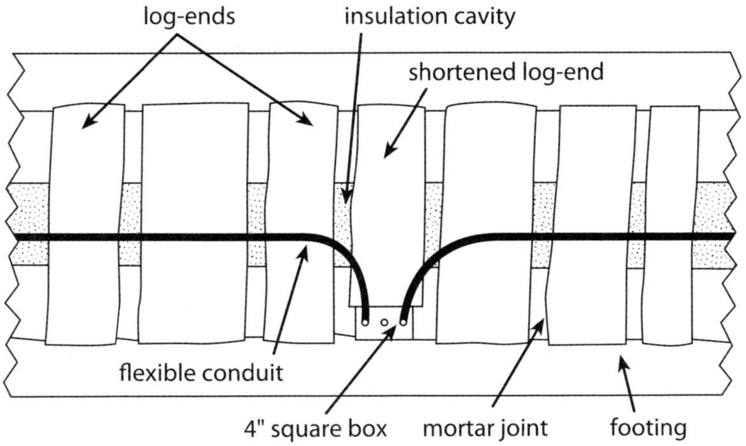

6.22: *Electrical boxes can be supplied by flexible conduit running within the insulation cavity.*

6.23: *Before laying the first course, we installed electrical boxes to the ends of 4-inch diameter by 8-inch long log-ends, and held them off the floor with 16-inch tall pieces of wood, so that the duplex receptacles (DRs) would be at the conventional height off the floor. These log-ends were later built into the cordwood wall. The Romex follows the insulated space between the inner and outer mortar joints. The little roofing nails sticking out of the plywood help lock the mortar to the floor. The wood is also treated with sealer and bonding agent prior to laying down the 4-inch-wide mortar joint.*

6.24: *The mortar has been laid for the first cordwood course, with the Romex cable running through the insulation space. The special log with the DR box attached is just lying there, waiting to be integrated into the masonry.*

6.25: *The wall is built up to and around the electrical box. Notice that plenty of cable is left sticking out of the box to make the connections to the electrical equipment easy for the installer.*

6.26: *We installed electrical boxes directly to the side of the 12-inch-wide door frame. The Romex coming up from the circuit breaker box on the first floor follows the insulated space, and is stapled between two 1-inch by 1-inch key pieces on the door frame. The M-I-M stick shows where the mortar, insulation, and mortar will go.*

6.27: *A view of the other side of the door frame shows the key pieces more clearly, as well as the first course of cordwood masonry.*

6.28: *A double-wide box fastened to a door frame (left side of image) will provide switches for an overhead light, as well as the LED light inside a crystal skull bottle-end feature.*

the Romex was always more than four inches from either surface of the wall. He brought the conductors into the back of his electrical boxes, which were set flush into large log-ends. During the winter prior to building, Ed prepared 20 or so 10-inch diameter logs for this purpose, by cutting and chiseling correctly sized rectangular openings into the logs to receive the boxes. He routed a pathway from the box opening to the center of the log to carry the Romex from the insulation cavity into the back of the electrical box.

When we built the new second story at our Mushwood Cottage, I took the opportunity to photograph how we integrated electric into the cordwood walls; this is shown in the photo essay of Figures 6.23 to 6.28.

There may be sections where it is not possible or desirable to hide the wire in the cordwood wall or under the floor. In those cases, wiring may be enclosed using Wiremold or electrical metal tubing (EMT) conduit on the interior cordwood wall surface or along posts, beams, and window or door frames. Exposed conduit or Wiremold is code approved and has several advantages for the cordwood masonry builder: Using this method, cordwood masonry production is not further slowed by taking time to weave conduit or Romex through the insulated cavity. Electrical can be installed after the cordwood walls are built and the roof installed. Also, the electrical circuits are readily accessible to facilitate changes, repairs, or additions.

There are some disadvantages to surface-mounted wiring. The Wiremold or EMT adds extra cost to the electrical component. New skills must be learned to make a nice job of surface-mounted wiring. And some people may not like to see surface-mounted conduit — although it is becoming more common all the time, particularly in commercial buildings. Wiremold (and other available systems) comes in a variety of colors, and EMT conduit can be painted to match or contrast. By careful planning and intersection with interior partitions (where conventional wiring practices may be used), it is possible to minimize the amount of surface-mounted wiring quite a bit, although the National Electrical Code (US) does require a duplex receptacle every 12 feet around the perimeter of all rooms.

Feeds may be run from the distribution panel to points around the building by using the space left between the inner and outer wooden plates, often made from 2-by-6-inch planking, at the top of the cordwood wall, if your construction method happens to incorporate that detail. Wiring to lighting fixtures can be run along the top side of girders, if exposed post-and-beam construction is used in the home.

Some builders have run a baseboard around the base of cordwood walls, incorporating conduit or Romex conductor behind the baseboard and surface-mounted boxes on the baseboard surface. If the first course of similarly dimensioned logs is cut an inch or two shorter than normal — 14 inches instead of 16 inches, for example — the baseboard need not protrude into the room.

Plumbing and mechanical systems with cordwood

Exhaust plumbing goes under the building, and interior plumbing is normally part of interior walls, so there are no special concerns with cordwood masonry, which is usually used for external walls. Similarly, mechanical systems — heating and air conditioning — are normally contained in the home's interior. One caveat to this would be to make sure that condensate from any external wall AC systems is channeled away from the cordwood wall.

Top left: *Son Darin Roy built this tree into the wall of his 20-sided cordwood home, Driftwood, next door to Earthwood.*

Center left: *Detail of a panel at "Driftwood," Darin Roy's cordwood and dome home, West Chazy, New York. Two blue geodes, a red cedar log-end, and a mushroom cubbyhole.*

Bottom left: *The "Easter Island" panel at Earthwood.*

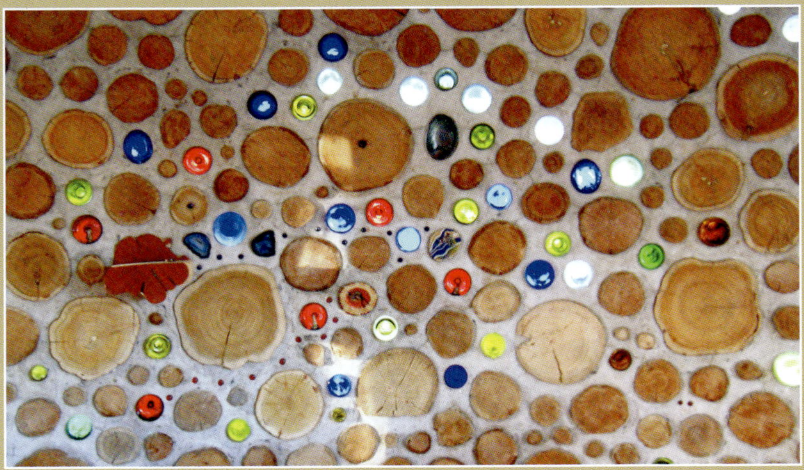

Center right: The Crystal Skull gets natural light in the day through a gallon jar on the exterior, and a switched LED light for night time use.

Bottom right: Our Australian panel incorporates souvenirs brought back from Oz.

Top: *Mushwood Cottage. Chateaugay Lake, New York, is a two-story round load-bearing cordwood building, 22 feet in diameter, with 12-inch thick walls.*

Early cordwood buildings at the author's Log End Homestead, 1975-80. Center right: Log End Cave, an earth-shelter with cordwood masonry on the southern exposure. Bottom right: Log End Sauna, the first of many 6-posters which have been built all over the United States, as saunas, guest houses, and garden sheds.

Bottom left: *"Littlewood" was originally built as a playhouse and school bus stop. Son Rohan, then 7, two other 7-year-old boys ,and a 12-year-old girl did the cordwood in 1983. It still stands.*

Top: *Earthwood Building School, West Chazy, New York. The main building is 2400 square feet, load-bearing round, 40 percent earth-sheltered with a full living roof. Built in 1981-2.*

Center left: *This 20-sided cordwood building home was built by Barbara Pryor and Steve Coley near Buena Vista, Virginia.*

Bottom left: *The Arcus Center for Social Justice Leadership, Kalamazoo College, Kalamazoo, Michigan, is a 10,000 square foot architect designed cordwood building. The image shows the building under construction. The complete story comprises Chapter 16 of* Cordwood Building *(New Society, 2016).*

CREDIT: STEVE HALL © HEDRICH BLESSING

The John Mecikalski General store (also a hotel) was built in 1899 and fully restored between 1985 and 1987, thanks to a grant from the Kohler Foundation .It is listed on both the Wisconsin and National Register of Historic Places.

Top and center right: *Nancy Dow laid almost all of the cordwood at Ravenwood, Saranac, New York. She keeps a very nice balance of log-end sizes and shapes.*

Bottom left: *Greg Hart, a student at a Wisconsin cordwood workshop, went home to Traverse City, Michigan, and made a barbeque wagon decorated with "cordwood paneling," thin log-end disks fastened to a plywood backing. Greg Hart photo.*

Bottom right: *This cherry log-end has become very hard, almost petrified, in the 35 years since it was laid up. Several checks formed, but the one at the bottom broke through and became the primary check, now stuffed with gray backer rod topped with clear caulk.*

Credit: Greg Hart

Top: *Demonstration "cobwood" wall built by Rob, Jaki, and volunteers at the Natural Building Colloquium, Kingston, New Mexico.*

Center left: *Recent image of exterior wall at Earthwood, built in 1981. Log-ends were made of old split cedar fence rails. They have weathered beautifully.*

Center right: *Cobwood panel (right), bordered by three panels made with a Portland cement-based mortar.*

Bottom left: *Log-ends with a non-existent mortar joint? No, this is a placemat made (in China!) of eighth-inch disks of hardwood stuck to a felt backing. But similar details have been used as interior walls in various bars and restaurants. Pretty nice balance of large and small rounds!*

Stoneview Guest House at Earthwood is a 22-foot diameter octagon (long diagonal). With 8-inch cordwood walls, this yields 256 SF of actual usable space, enough to house four guests with its own bathroom.

Center right: *Interior of "The Cordstead," built by Sandy Clidaras at Ste. Ann des Lacs, just north of Montreal, Quebec. It is available as an Air B n B. Go to http://thecordstead.blogspot.com*

Bottom left: *Jaki and Rob Roy at their round Mushwood Cottage on Chateaugay Lake, northern New York. Many of the larger interior log-ends were sanded and treated with two coats of water-based urethane.*

Bottom right: *Interior of Ravenwood, Bruce Kilgore and Nancy Dow's large two-story cordwood triangular (Trisol) earth-shelter. Walls are double-wall cordwood, 26" thick. They heat their northern New York house with about two cords of wood per winter.*

Top left: *This little bird has found a permanent home in Bruce and Nancy's wall at Ravenwood, Saranac, New York.*

Top right: *A Scottish sheep farmer — and close friend — carved this Celtic dragon from a piece of cherry turned on a lathe. The piece is 1.5 inches thick, so we cut a 12-inch long log-end to 10.5 inches and pointed up to the feature. Mushwood Cottage, Chateaugay Lake, New York.*

Center left: *Shelves can be made of a log split down the middle. (The faux ivory carving is only an inch thick and mortared up to a log-end of the same diameter, cut a little short).*

Bottom left: *Got a favorite statue? You can build it into either side of a cordwood wall.*

Bottom center: *Mushrooms are always popular and — usually — easy to do. Tom Huber (this design) and Darin Roy (first colour page, center left) are both strong mushroom aficionados in real life.*

Bottom right: *One of our Canadian students did nothing but cordwood design features for five days. He couldn't resist immortalizing his instructors into our sauna wall. Rob is the frowning figure, Jaki the smiling cutie pie.*

Top left: *A student builds cordwood as Jaki Roy looks on, near New Norfolk, Tasmania, Australia.*

Top right: *Jaki and I wrapped "bull-nose" corners around the outside of the corner posts at a building done during a cordwood workshop at the La'akea Community, near Pahoa, island of Hawaii.*

Center left: *Here, I am laying a large log-end. Note the red plastic level, a valuable aid to assist in proper placement.*

Center right: *Jaki Roy points a wall made with cob (clay) mortar at a workshop in Virginia.*

Bottom right: *The easiest way to cut bottles, far and away, is with a slate and tile cutter.*

Bottom left: *This special bottle, in the shape of a fish, is matched with three clear glass cylinders made from jars or clear cut bottles.*

Chapter 7

Design Notes for Cordwood

A DESIGNER OF CORDWOOD MASONRY BUILDINGS will need to choose from the options presented in the previous chapter in order to create a building that functions as an integrated system. In addition, the various points given in this chapter should be considered.

1) *Foundation*

a) Cordwood masonry is most often — and safely — placed on a concrete foundation, engineered to support the load.

b) Stone foundations have proven the test of time on cordwood buildings 100 to 150 years old.

c) Small buildings, such as sheds, have been built successfully on old railway ties.

d) Provide a reasonable space — four inches — between the top of the foundation and the beginning of the cordwood wall. This prevents long-term contact with ground moisture. A row of solid concrete coping blocks on the foundation can provide adequate clearance. Snow standing against a cordwood wall has not caused a problem in any case of which the author is aware.

e) If building on a monolithic floating slab, use a timber frame and roof to provide umbrella protection so that water does not collect on the slab and stand against the first course of cordwood.

f) In areas prone to ground termite infestation, incorporate a termite shield around the edge of the foundation. Provide effective flashing/sealing detail at wall/foundation intersection.

2) *Overall wall width*

a) Cordwood masonry wall thickness should be designed to meet the thermal performance targets consistent with the kind of building it is: house, garage, shed, sauna, etc.

b) Double-wall cordwood masonry, as described in this book, can provide exceptional thermal performance in even the coldest Canadian climate, but it is a very time-consuming process.

3) *Exterior finishes and wall protection*

a) Specify roof overhang and gutters appropriate for climate to protect walls from excessive precipitation.

b) Exterior cordwood masonry will gradually take on a pleasing weathered appearance. It is a labor-intensive process to treat log-ends in such a way so as to diminish this natural weathering.

c) Sanding the exterior of log-ends and applying Cabot's Siliconized Sealer will lessen the weathering effect and provide some protection against wind-driven rains. To maintain this protection, it may be necessary to reapply the sealer every 4 to 6 years.

4) *Interior finishes*

a) Interior cordwood walls will keep their original appearance.

b) The author has brightened interior walls and/or brought out color in special-feature log-ends by sanding them with a circular sander and applying two coats of a water-based urethane.

c) The interior of cordwood walls can be plastered (rendered).

d) To brighten a dark area, the interior of cordwood walls can be painted with a 50/50 mix of ivory latex paint and white joint compound.

5) *Window and door details*

a) Doors should be framed with heavy timbers, such as 4-by-8-inch stock (for an 8-inch wall), doubled or tripled to make up the width of 16-inch or 24-inch cordwood walls.

b) Windows, whether fixed or opening units, should be surrounded in heavy "window bucks," usually made out of full-sized 2-by-8-inch stock (for an 8-inch wall), doubled or tripled to make up the width of 16-inch or 24-inch cordwood walls.

6) *Electrical details*

a) Electrical boxes should be connected to framing members or set into a space in a large log-end created for the purpose.

b) Insulated electrical wiring should never be in contact with the cordwood mortar, which can cause damage to the insulation. If necessary for a particular detail, sheathe the wiring in conduit in a manner that meets local electrical codes for wiring in contact with concrete.

c) Wiring can run inside the insulated space in a cordwood wall providing it is at least four inches back from both inner and outer wall surfaces. Otherwise, run the Romex cable — or its equivalent — inside of conduit.

7) *Top of wall*

a) When using cordwood as infill with a strong timber frame, roof framing can be fastened to the structural frame, the girts.

b) Ground floor joists can rest on the inner edge of foundation footers if a damp-proof course is installed to prevent rising damp.

c) Second floor joists can rest on girts or be hung on girts with correctly sized U-shaped mechanical fasteners made for the purpose.

Chapter 8

Construction Procedure

Site Preparation

For mixing mortar. Have your constituent ingredients — sand, lime, Portland cement, wet sawdust (if used) — all within a shovel's length of the wheelbarrow, which should be on flat ground. I keep my bags of Portland and lime up off the ground on a pallet, and cover them with an upside-down wheelbarrow at the end of the day. Have water on site: a hose, if possible, or a drum of water with buckets. Have mortar pans ready to load.

For making sawdust insulation. I get my sawdust from a local sawmill with my pickup truck, still for $5 in 2017. I load the truck with a snow shovel, and unload it the same way on site, placing it conveniently in the mixing area, near to the lime. For insulation, you can use hardwood or softwood sawdust, but softwood generally gives a better R-value. (Do not use dense hardwood sawdust for the mortar additive; it doesn't work.) A half-inch screen (Figure 8.5) and an old clunker wheelbarrow are useful.

For laying cordwood. The concrete, solid block, or stone foundation should be swept clean and dampened. Brush a liquid bonding agent onto the parts where you will be placing mortar that day. Several companies make a bonding agent for this purpose: DAP, Thoro Corporation (Acryl-60), and others. (You can begin to lay the mortar even if the bonding agent is not perfectly dry, but it needs to be renewed if left exposed overnight.) Make sure your prepared cordwood is standing on end within an arm's reach of where you will be building, as per Figure 8.1.

8.1: *The area to be worked on this 8-inch cordwood wall (for a sauna) has had a bonding agent applied to the concrete foundation where the cordwood will be laid. A variety of 8-inch log-ends stand within reach.*

Making Mortar with Wet Sawdust

When making mortar with a wet sawdust as the additive, it is imperative that the sawdust be passed through a *quarter-inch* screen and thoroughly soaked at least overnight in a non-leaking vessel, such as an open-topped steel or plastic drum, old bathtub, etc. So, the last thing to do each day is to make sure that enough sawdust is soaking for the next day's work.

Adding materials to the wheelbarrow

Review "Mortar Materials," Part 2 of Chapter 5.

We add the ingredients to the barrow by the shovelful, using the following cadence, which greatly reduces mixing time. For the Portland mix (same recipe given in Chapter 5):

- 3 shovels sand — 1 shovel sawdust — 1 shovel lime — 1 shovel Portland
- 3 shovels sand — 1 shovel sawdust — 1 shovel lime — 1 shovel Portland
- 3 shovels sand — 1 shovel sawdust — 1 shovel lime

The introduction of the constituent ingredients in this manner places the Portland cement one-third and two-thirds of the way into the mix. As only two shovels of Portland are used, there is none in the third line. Using this cadence reduces time needed on the "dry mix." Make sure that the shovelfuls are equal-sized for all ingredients. Use a sturdy industrial barrow, not a flimsy garden type. (We're still using the same two metal wheelbarrows that we used at Earthwood in 1981. But, at the end of each day, we have always been careful about cleaning them with a scrub brush and doing a final rinse.) Use two spade-type shovels of the same size, one for the dry goods, one for the wet sawdust. Don't put the "wet" shovel into the cement or lime bags; this soon makes a mess out of the shovel.

Place the cement and lime carefully in the barrow. Do not throw it. If you get cement in your eyes, do a 15-minute rinse with clean tap water and seek medical attention. A combination respiratory mask with goggles is an added protection when handling the dry powders.

With the masonry cement mix — again, see Chapter 5 — a good cadence for adding material is:

- 3 shovels sand — 1 shovel sawdust — 1 shovel masonry cement — 1 shovel lime
- 3 shovels sand — 1 shovel sawdust — 1 shovel masonry cement — 1 shovel lime
- 3 shovels sand — 1 shovel sawdust — 1 shovel masonry cement

The numbers in these mixes refer to equal parts by volume, so always use the same size of shovel and load it the same way each time — small, medium, or heaping — depending on the size of batch you want. Tip: A little wiggle of the shovel yields consistent, medium-sized shovelfuls, and makes a nice wheelbarrow load.

Use strong cloth-lined rubber gloves throughout the project, including during the mixing process. Wet cement will eat nasty little holes in your hands (discussed in Chapter 12, "Rubber Gloves and Facemask").

In an industrial-strength wheelbarrow, dry-mix the goods with an ordinary garden hoe until the mix is a uniform color. Then make a little crater in the center and add water. How much water to add depends on how wet the sand and sawdust is. I remember one occasion when the sand was so wet that I didn't have to add water at all! For the first batch of the day, go easy on the initial splash of water, say a quart (liter) or two. Mix it thoroughly and conduct "the snowball test." Toss a snowball-sized glob of mortar three feet in the air — one meter in Canada — and catch it in your gloved hand. If it shatters, it is too dry. If it goes "sploot!" like a fresh cow pie,

8.2: *After introducing the materials into the wheelbarrow, I dry-mix it with a garden hoe, pulling it in each direction two or three times.*

8.3: *After everything is a consistent color — remember to scrape up the sand at the bottom of the barrow—make a crater of water and commence wet mixing, back and forth with the hoe until the consistency is right.*

it is too wet. If it holds its shape, doesn't crack or crumble, and is nice and plastic (cohesive), it is just right. (Note to experienced masons: You want *stone* mortar, not brick or block mortar. You folks know the difference.)

If the mortar is too dry, add more water, remix, and test it again until it is plastic and cohesive; if too wet, you can add a little more dry goods — but in the same proportions — until it is right. Do not add more soaked sawdust if the mix is really soupy, or you might have trouble getting it stiff enough.

If you are working alone, or with a partner, you can actually wheel your mortar to site and work right out of the barrow. Down low, though, on the first couple of courses, it is easier to load a mortar pan in order to bring it closer to hand for efficiency. This reduces frequent getting up and down, nice over the age of 60.

8.4: *Test frequently until you know your mortar. Do the "snowball test."*

Making Mortar with Cement Retarder

First, review again the Sidebar in Chapter 5, "Commercial Cement Retarders."

The mortar mix is slightly different when using a retarder: an extra part of sand replaces the three parts of soaked sawdust. So, with Portland, the recipe is 10 sand, 2 Portland, 3 lime. With masonry cement, it is 10 sand, 3 masonry cement, 2 lime.

With a Portland mix, a good cadence for introducing the materials to the wheelbarrow would be:

- 3 shovels sand — 1 shovel lime — 1 shovel Portland
- 4 shovels sand — 1 shovel lime — 1 shovel Portland
- 3 shovels sand — 1 shovel lime

8.5: *I shake the sawdust through a half-inch screen, removing various matter that I don't want in my insulation.*

With masonry cement:

- 3 shovels sand — 1 shovel masonry cement — 1 shovel lime
- 4 shovels sand — 1 shovel masonry cement — 1 shovel lime
- 3 shovels sand — 1 shovel masonry cement

Proceed as above for "Making Mortar with Wet Sawdust." After the dry mix is complete, make a crater in the center of the wheelbarrow and pour about a half gallon of water into it. It will look like Figure 8.3. Now add the cement retarder to the water (three ounces with the ones named in the Chapter 5 sidebar about cement retarders) and stir it around. Do the wet mix, adding water as necessary to achieve the same stiffness and plasticity described for the sawdust mix. Do the same "snowball test."

Before your project actually begins, do an experimental batch — with cordwood — and watch it for five days. If there has been no mortar cracking, continue in the same way. Sometimes you have to try a little more or a little less retarder to get a favorable result. (Testing the sawdust mix is a good idea, too.)

Making the Sawdust/Lime Insulation

We pass the sawdust through a half-inch screen (Figures 8.5 and 12.6), which catches bark, chunks of wood, and other detritus that inhibits easy pouring into the cavity. To prevent vermin, we treat our sawdust with builder's lime at the ratio of 12 parts sawdust to 1 part lime. Also, if the sawdust gets wet from rain during (or after) construction, the lime will set up with the sawdust, dry it out, prevent mold, and leave an effective bead board type of insulation product in the cavity. About 25 years after building Earthwood, we had occasion to take out a panel of cordwood in order to expand a window space into a door. The sawdust insulation was just as good as the day it was installed.

We mix the sawdust and lime in a wheelbarrow. In this case, it can be either an old clunker or a lightweight garden barrow. A mixing tip: screen six shovelfuls of sawdust into the barrow, shaking it through the half-inch mesh screen. Add one shovelful of lime. Now top this with six more screened shovels of sawdust. The lime is in the middle of the batch, like the cream filling of the famous cupcake. Mix with your hoe until the entire batch has the same color throughout. If the sawdust is too dry, spray a little water in as you mix, so that the finished product does not blow all over your mortar during or after installation. Plus, you're less likely to inhale the lime dust. A good dust mask or respirator is advisable when shoveling dry lime — or cement — whether you're mixing sawdust insulation or mortar. Store your sawdust/lime insulation under cover.

Before Beginning the Cordwood: Key Pieces

The examples illustrated in this chapter show 8-inch sauna walls being built within a small, strong timber frame. When building within a timber frame, lock the cordwood masonry panels to the surrounding timbers with a mechanical key, particularly the vertical posts, but also up against the girt (plate beam) on longer panels. Similarly, door and window frames are keyed to the cordwood masonry, as will be seen.

Positive wooden key pieces are screwed or nailed to the edge of the posts against which the cordwood masonry will bear. They are made from full one-inch-thick boards, their width being the same width as the cordwood wall's insulated space. For the project shown in this chapter, a sauna with 8-inch walls, the key pieces are made from 1-by-2.5-inch boards, because the insulated space is 2.5 inches wide, as per Figure 12.7. You can make it of short scraps; you don't need full post-height boards.

When laying cordwood, it is okay for log-ends to actually touch the key pieces, which, in this case, are attached to the middle vertical third of the post. One inch is a perfect mortar joint. When fully set, the mortar matrix is mechanically locked to the post, because of the vertical wooden key. If this were a small horse barn, and the horse wanted to kick the wall out with his hooves, he would have to break the mortar matrix to do so, and it would still be difficult. With no locking key, and no mortar-to-wood chemical bond to the post, the horse could easily kick the wall out. Besides bucking horse protection, the key pieces give support during seismic events. A key piece can be seen in Figures 8.7, 8.8, and 8.10.

Building the Cordwood Wall

Order of events. There are three components to a cordwood wall. In order, they are: mortar, insulation, and wood (log-ends.) And that is the only order in which they are placed — on the foundation or on a wall in progress. Seems simple. Yet we constantly observe students, particularly during the first two days of a workshop, getting this out of order, looking for a log-end before the insulation is down ... sometimes, even before the mortar has been laid! The log-ends are the feature of the wall, and people can't wait to select them. Get the mortar down first! Then the insulation! Then let the wall build itself. *It* will tell *you* which log-end you need. Cordwood masonry is already a relatively slow, labor-intensive process. Don't waste time. Remember the mantra: "Mortar. Insulation. Wood."

Placing mortar. To lay the mud, grab a gloved handful from the barrow or mortar pan and place it on the foundation (already coated with bonding agent). In the example shown, an 8-inch-thick cordwood wall, you can try both a single-handed method or a two-handed method.

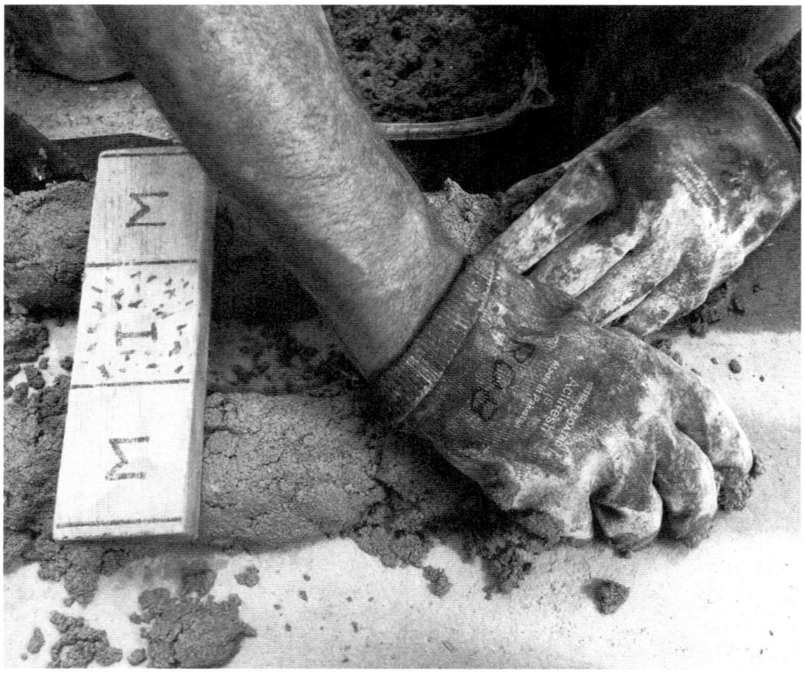

8.6: *With the faster single-handed method, my right hand places the mortar on the prepared place for the inner mortar bed, while my left hand does the outer bed. The M-I-M-stick helps me gauge width.*

8.7: *The two-handed method works well with wider mortar joints, but people with small hands might like it on the 8-inch wall, too. Note the mortar pan in the picture, and the sawdust bucket, both close at hand.*

The mortar joints are only 2.5 inches wide, and I find that I can grab a handful with each hand and place them down at the right width and at the same time, one on the interior, one on the exterior, as per Figure 8.6. With wider mortar joints, 4 inches or better, I use two hands to place one glob of mortar, as per Figure 8.7. In either case, shape the mortar so that it takes on the required cross-section shown in Figure 8.9. Do not overwork the mortar. We tell our students, "No pitty-patting!" Just place it and move on. Pitty-patting is worse than a waste of time. It actually brings moisture to the surface and promotes rapid setting of the mortar. People do it because they are trying to make their mortar bed look nice. Don't! You'll make it look nice during the pointing process.

How long you make each new section of mortar is a function of experience. When you are first learning, a length like that in Figure 8.1 — about two feet — is a good start. After a while, you will be comfortable running a double mortar joint out three, or even four, feet. It is important that you do not put down more mortar than you can comfortably cover with wood before it begins to stiffen. This varies with the day. With cool damp days, you have a lot more working time than on hot dry days, maybe two or three times more.

Placing insulation. Once the inner and outer mortar joints are installed, you can install the sawdust insulation. (Other insulation choices are given in Chapter 5.) With a small plastic spouted bucket, use its spout to direct the sawdust evenly along the length of the gap between the mortar beds. Do not hold the bucket so that its lip is perpendicular to the run of the wall. It will most assuredly spill onto your mortar. But, if it does, scrape the dry sawdust off as best you can, and apply a little more mortar to assure a good continuous mortar joint. See Figures 8.8 and 8.9.

8.8: *Pour the insulation between the inner and outer mortar joints with a small spouted bucket, being careful not to spill any on the mortar.*

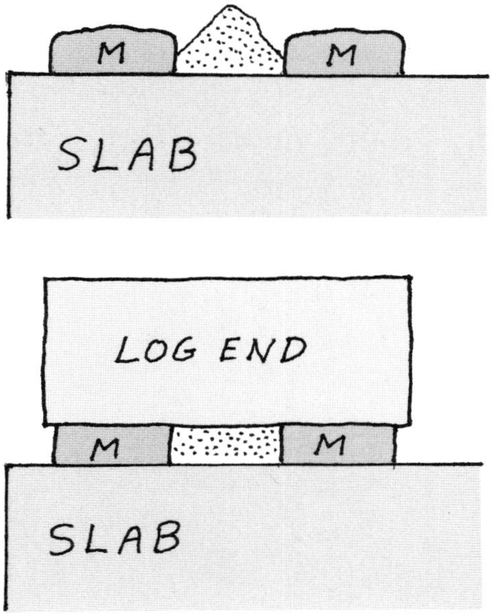

8.9: *Quickly make a roughly rectangular cross-sectional shape with the mortar. When both mortar beds are in place, pour in the sawdust insulation, leaving it a bit high in the middle. Bottom: The log-end will push the sawdust into the voids either side of the raised ridge.*

Placing the log-ends, first course. Examine your intended log-end. If there is a brighter or cleaner end, place that end to the interior. (The exterior is going to weather anyway.) If you have an obvious primary check on a round log-end — a check that goes right through the log — place the log so that the check is somewhere between 4 o'clock and 8 o'clock, so that it does not hold and distribute water from a driving rain.

Place the first log-end onto the mortar, an inch away from any post or previously laid log-end. Set it with a slight back and forth vibration, creating a suction bond. If you have a variety of sizes, as per Figure 8.1, set subsequent log-ends randomly with respect to size. This sends the wall into a pleasing *random rubble* style, and the wall begins to "build itself." Exception: if all your log-ends are the same size, simply lay them in regular courses, like a section of a beehive, still leaving an inch between log-ends.

Subsequent courses. Place the mortar by following the hills and valleys created by the random placement of the logs of the first course, maintaining the correct width according to the M-I-M-stick, and the usual one-inch thickness around the log-end. Next, pour in the insulation as before, slightly tamping it between log-ends with the handle of your pointing knife. Near a post, it is easier — and tidier — to place insulation with a small tin can instead of the bucket.

The mortar joint, seen from the side, will have variously sized "cradles," our name for the valleys. *Now* is the time to look for a log-end! Take a mental picture of the size and shape of the cradle, then look at your pool of variously sized log-ends standing nearby. With a little practice, you will select the right log on the first try, almost every time. If the one you try doesn't fit down into the space, try a smaller one. If it rolls around in the cradle, choose a larger one.

Two caveats: (1) Avoid the "three-to-four inch trap" — a space of that size between logs — as you may not have a log small enough to put in that space, resulting in a three- or four-inch-wide mortar space. This is more of an esthetic error than a structural problem. Slightly larger log-ends can prevent this trap, or smaller ones that allow yet another very small one between. (2) If you notice

8.10 A

8.10 B

8.10 C

8.11 A

8.10: *First course. (A) The first two log-ends are laid, a six-incher next to the right-hand post and a small one to its left. (B) A larger eight-incher has been placed up against an inch of mortar next to the small one. (C) Mortar has been placed up against the eight-incher and against the post. A five-incher fills the space perfectly. Note that primary checks are placed facing downward.*

8.11: *Second course. (A) An inch-thick "cradle" is made with mud over the small log, and between the two larger ones. (B) A three-incher fills the cradle nicely, maintaining the one-inch spacing between logs. (C) The second course is nearly finished. One more cradle needs to be completed, toward the right of the picture, where mortar is exposed above the first course, but has not been installed against the two small log-ends. Then the last log-end can be placed.*

the wall taking on a pattern of similarly sized log-ends — it looks like a section of a beehive — do something deliberate to break out of this pattern, such as placing a larger or smaller log. If you don't do this, you will lose the random rubble pattern and you will soon deplete a certain size of log-end.

8.11 B

8.11 C

8.12: *Here's the same small panel, up about four courses. Note the "random rubble" style, a good balance of small, medium, and large log-ends. The bark on the eight-incher (Figure 8.10) was deliberately left on, as it was thin and was obviously not going to come loose because it had been drying two years already. We chamfered the edges with our scraper, so that the bark is no longer apparent in this image.*

Laying large log-ends. Large log-ends — 10 to 12 inches in diameter or greater — need a little planning. Sometimes a perfect place presents itself, but not often. Also, you may want to feature a special big one at a certain spot in the wall. If a space or cradle is too big for a particular log — it rolls around — put one or two smaller ones in, right or left, to reduce the size of the

8.13: *Setting a big one. (A) A likely spot presents itself in the wall, and it's about time for a big one; in this case, 13 inches in diameter. No big ones? An alternative here would be to place a little mound of mortar just over the log-end below, breaking the space up nicely for two medium logs. (B) I try the big one different ways — seldom is a log-end a true cylinder — and find a way that it fits while maintaining a constant 1-inch-thick mortar joint. It's a very protected place, but I still try to keep the primary check in the log's lower half.*

8.14: *(A) A large log is a "quality control" opportunity. I check that it is plumb and level, using the plumb bubble of my 4-foot level perpendicular to the run of the wall. (B) A small level across the top is useful, as well.*

space to what is needed. If the space is too small, first try turning the log, as some parts of it have a tighter curvature than others. If that doesn't work, you might have to build up the bottom of the cradle with smaller log-ends … or wait for a better opportunity. Try to balance the large logs in the panel as a whole.

Keeping straight walls … straight! When building within a timber frame, it is fairly easy to keep the wall going up straight by "eyeballing" from post to post to see if errant log-ends are either sticking out or have not been built out far enough. Keep in mind that some logs might be a little longer or shorter than the ideal. You can either "average" the error, or you can choose to keep one side of the wall in the same flat plane, usually the interior surface. Today's log-ends (not yesterday's!) can be moved in or out by a sharp rap with a hammer. A fist tends to move the whole wall.

Another way to keep the wall straight is to stretch a mason's line from one post to the other. The wooden or plastic clip keeps the line one-half inch away from the desired plane. When laying log-ends, always lay them so that they are about a half-inch from the line. If you touch the line with the log, it will no longer be straight. This method works well with stackwall-cornered cordwood construction, too.

8.15: *(A) The author eyeballs from post to post in search of errant log-ends. He spots a small one sticking out too far. (B) The sharp impact of a hammer will allow log-ends to be brought back into line.*

8.16: *This wall* may *or may* not *have been built using a mason's clip and line like this one. But* it *could* have *been.*

8.17: *Mason's clips can be hard to find — and hard to work with if the corner is not perfectly square. You can tack a nylon line onto a post as shown below and hold it proud of the wall with a uniform (¼" or ½") wooden shim.*

Keeping a round building's walls vertical. With round buildings where the cordwood is load-bearing, clip one end of a tape measure to a nail or pin at the exact center of the building. Then, using the correct radius for the building — 15 feet, for example — use an indelible marker to draw the inner and outer wall circumferences on the foundation. Then, as you build, frequent examination of the plumb bubble of a 4-foot or 6-foot level will keep the wall going up vertically and round. Check after every six or eight log-ends, particularly the larger ones, which serve as good benchmarks. There is another way — dating back to castle-building days — that is more work initially, but will assure quality control, especially on buildings of 24 feet in diameter or greater:

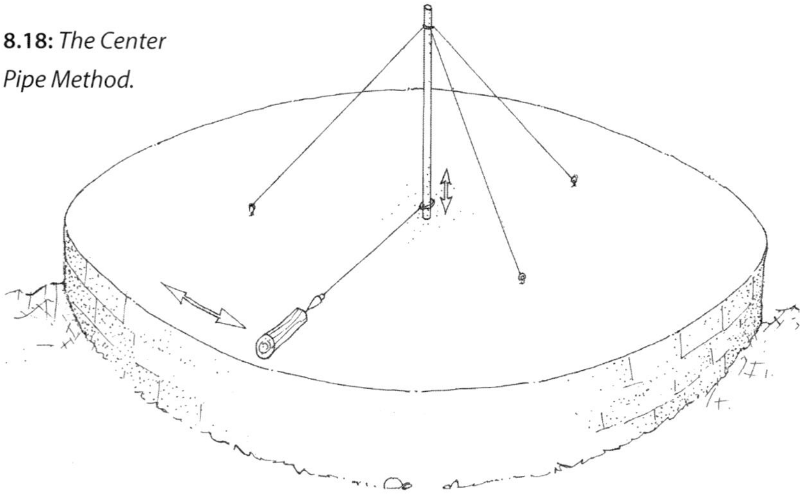

8.18: *The Center Pipe Method.*

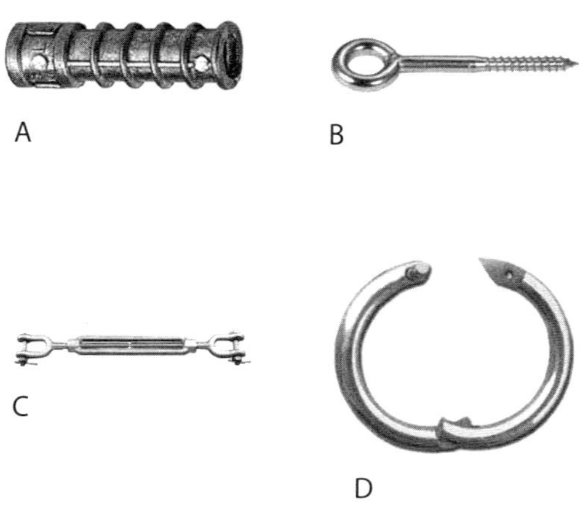

8.19: *Equipment needed for the center pipe method: leaded expansion shields to insert into slab (A); eye bolt to go into the expansion shield (B); turnbuckle for tensioning the guy wires (C); opening O-ring to slide up and down the center pole (D).*

Center pipe method

At the time of pouring concrete, place a short one-inch diameter pipe with a threaded end at the very center of the building. In the slab, drill three half-inch holes, two inches deep, into the slab, each at about six feet from the center and equally spaced every 120 degrees around the circle. Place leaded expansion shields into the holes. When ready to build your round cordwood wall, place a female union fitting onto the pipe. To this, insert a pipe as high as the wall you want to build, seven feet for example. Predrill quarter-inch holes all the way through the pipe, every 9 inches along its length, so that a 16-penny nail can be inserted through at various height intervals. In the three expansion shields, screw in O-ring screws of the correct diameter for the expansion shields. Now plumb the pipe perfectly vertical, with cables stretched from its top to the three O-rings sticking out of the slab. You can plumb and tighten the three cables with turnbuckles.

Place a bull's ring around the pipe so that it can easily slide up and down. Tie a non-stretch line to the bull's ring and put a plumb bob at the other end, so that, when extended horizontally, the total length of the line from the center of the building to the tip of the plumb bob is exactly the same as the inner radius of the circle. Now build in courses. Lay the inner surface of each log-end up to the plumb bob at the end of the extended line, assuring that each is equidistant from the center and that the course is round. To keep the line horizontal — and the wall plumb — move the bull's ring up as needed and support it by a 16-penny nail. Now your wall will be curved correctly, as well as plumb. Thanks to Pythagorean trigonometry, it is not imperative that the line be perfectly level; within five degrees is fine. But if you don't move the bull's ring up at all, your building, after a while, will start to look more like a dome than a cylinder.

We have seen this system used on a 32-foot diameter home, and it worked very well. But we have had good results with the simpler plumb bubble method. The choice is yours.

Pointing

Pointing, also known as "tuck-pointing" or "grouting," is a critically important part of cordwood masonry, and it accomplishes several purposes.

First, good, stiff pointing maximizes the friction bond between wood and mortar. Remember that there is no chemical bond between the two, so a good friction bond is imperative.

Second, pointing beautifies the wall. Jaki can point a poorly laid wall and make it look better than a well-laid wall that is not pointed or poorly pointed. Do both: lay it up well and point it well. I have seen the opposite: a badly laid wall, poorly pointed. This gives cordwood masonry a bad name.

Third, good pointing smoothens the mortar, making a more water-repelling surface on the outside, as well as a less-dusty interior.

Fourth, if the pointing is recessed slightly, say one quarter to half an inch, and all or most of the log-ends in the wall shrink, it will be easy to conduct a repair. (See Chapter 10.) Recessed pointing also looks better. The log-ends are the defining feature of a cordwood wall. Having them stand proud of the mortar is what gives the wall a pleasing texture, similar to the relief found in good stone masonry.

You'll need a few pointing knives, described in Chapter 12.

Jaki, queen of the pointers, does a "rough pointing" first, using just her rubber gloves. She removes excess mortar and catches it in her

8.20: *Jaki rough points the wall, using just her rubber gloves, assuring that the "reveal" will be the same throughout the panel, about a quarter inch.*

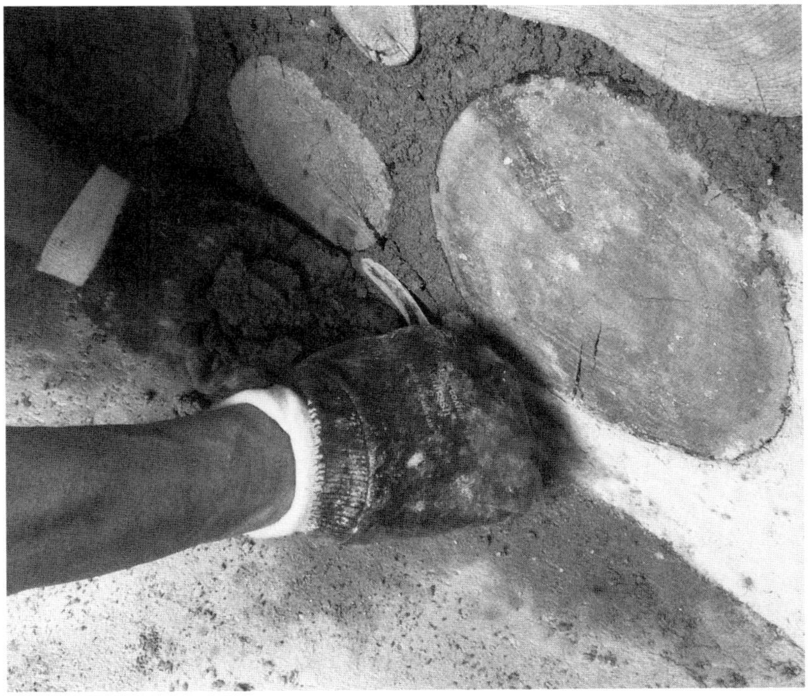

8.21: *Rob rough points a little differently. He cuts away excess mortar with his knife, catching it in his gloved hand. Then he pushes this mortar into any needy cavities. He, too, looks for a consistent quarter-inch reveal. Try both methods to see which works best for you.*

gloved hand. Then she uses her knife to press it off her hand into any gaps. "Borrow from Peter to pay Paul," she says — not good economics, perhaps, but it works with cordwood pointing.

For the finished pointing, press quite stiffly with the knife blade while drawing it along the mortar joint. Draw the mortar out smooth, removing knife marks. How meticulous you want to be is up to you, but keep a consistency of style. *Do not over-point.* You can be so fussy, going over and over the work, that you will simply bring a lot of water to the surface, which will cause cracking of the mortar within a few days. We've done it.

Do not finish-point the wall too early or too late. How do you know what the right timing is? It is too early if the mortar is slumping from a log or if little air bubbles are forming. (Slumping could also be a sign that the mud was mixed too wet.) It is too late if you can't take irregularities out easily. Exactly when to point will depend on the drying conditions for that day. In hot dry conditions, you may have to finish-point within an hour. Damp cool days may allow pointing two or three hours later.

8.22: *I draw the knife along the mortar joint with a fairly quick motion and with some pressure, smoothening the mortar. I take my glove off my pointing hand only, for better "feel" of the knife, but I am careful not to touch the mortar. Jaki points beautifully with her glove on.*

Some builders like the slightly rougher texture of "brush pointing." Do a good rough pointing with the gloves, maximizing the bond, then go over the work by drawing a three-quarter-inch paint brush over the pointing.

Making Window Bucks

A "window buck" is a heavy frame built into the cordwood wall to accommodate whatever windows you want to use, be they opening types or fixed thermal pane units. Cordwood masonry is very accommodating in allowing freedom of window choice. You can go to a local manufacturer of thermal pane glass and buy units that were unsold for some reason (maybe they were cut the wrong size); often you can procure perfectly good units for 10 or 20 cents on the dollar. Likewise, you can pick up windows that were removed from a building during remodeling. Jaki and I have found lots of perfectly good windows at roadside garage sales.

With cordwood masonry, I always make window bucks out of full 2-by material: 2-by-8 inch planks, for example, with an 8-inch wall. Or 2-by-12 inch stock for a 12-inch wall. At Earthwood, with its 16-inch walls, we made double-wide window bucks from two individual 2-by-8-inch bucks scabbed together with vertical 1-by-6-inch key pieces on the right and left sides, which also mechanically locked the buck into the cordwood masonry. (Do not install key pieces on the top or bottom; they are unnecessary there and make installation more difficult.)

For thermal pane units, make the rough opening (inner dimensions) of the window buck a half-inch greater than the dimensions of the window unit itself, both in width and in height.

So, a thermal pane actually measuring 30 inches wide by 40 inches high will require a window buck with an internal space of 30.5 inches by 40.5 inches. For this example, you will need two uprights of 40.5 inches, plus a top and bottom piece each measuring 34.5 inches (the 30.5-inch rough opening width plus 4 inches for the two laps of the top and bottom pieces over the side pieces). Nail or screw the top and bottom pieces into the side pieces. Finally, and importantly, square the buck perfectly with a framing square and fasten a wooden diagonal as shown in Figures 8.23 and 8.24. Try the unit in the buck, to make sure it fits, then store the unit safely — and vertically — so that it does not lose its airtight seal. (A neighbor lost the seals in about a dozen beautiful large thermal pane units that he stored horizontally, causing them to go cloudy.)

Installing Window Bucks in the Wall

When building within a timber frame, it is often possible to fasten the window bucks to the underside of sidewall or endwall plate beams (girts), or even against posts. Large windows

8.23: *This window buck is made of full-size 2-by-8 inch lumber. The heavy 6-by-8 inch sill and lintel were an optional design feature which the owner requested. The entire unit was leveled and plumbed, then vertical braces held it plumb against one of the rafters. Note the diagonal brace to hold the buck square, and the key piece on the side.*

8.24: *The bracing to hold the buck vertical is seen clearly. The buck is a good place to store log-ends and mortar pans during construction. Note also the good random rubble placement of log-ends.*

might fill the space between vertical posts or uprights, as we did at Log End Cottage and Stoneview. In these cases, simply fasten the window buck to the frame with structural screws, and, after checking for level and plumb, hold the buck squarely in place temporarily by screwing a diagonal wooden brace from the buck to a post and/or girt.

Often, particularly with curved-wall and stackwall-cornered buildings, it is necessary to "float" a window buck somewhere in the cordwood wall itself, like the unit shown in Figures 8.23 and 8.24. This is not difficult to do if you follow these steps:

1. From your elevation plan, determine the height of the cordwood wall to the underside of the window buck. (For appearance sake, common practice is to keep the tops of the windows at the same height.)
2. Make a little "idiot stick" out of a common wall stud or piece of straight scrap lumber. Measuring from the bottom of the stick, draw the location of the actual window buck. Then, an inch below that, make another line, which I call the "no-wood-higher-than" (NWHT) line. Draw a log-end on the stick, if you like, right up to this line.
3. Now, build your cordwood up from the foundation in the regular way, but, as you start to get closer to the NWHT line, you will need to influence the wall — which has been happily building itself — by planning log-end sizes to create a *mesa,* a series of log-ends with all of their tops corresponding to the NWHT line. This mesa needs to be at least two inches longer than the length of the window buck.
4. If it still early in the day, you can place long mortar joints and insulation upon this mesa — it takes quite a bit of mud — and set the window buck on it like a giant rectilinear log-end. Check it for level and plumb. If all is well, temporarily install scrap wooden bracing from the buck to the girt (or nearby post) to hold it in place. No frame to tie to? Drive a stout stake in the ground and hold the window buck plumb and level with a long diagonal brace screwed to the stake. If it is late in the day, just leave the tops of the mesa uncovered by mortar. Install the window buck the next day, giving the masonry a chance to set before loading it with fresh mortar and the buck. After installing the buck, it is a good idea to build up against both sides of it with cordwood masonry, to stop it from getting displaced by "things that go bump in the night." Remember that a vertical key piece on the right and left sides of the window buck locks it to the masonry.
5. A final tip: I have found it useful to make my mortar just a wee bit less stiff than usual, more like brick or block mortar. I'll put it on just over an inch thick. Then, with a four-foot level as my guide, I can tap the top of the buck to set it level and plumb, so that I have the desired one-inch mortar joint. A level is a necessary tool for all aspects of cordwood work, especially when installing door jambs and window bucks.

Installing the Windows Themselves

Thermal panes. Later on, when it comes time to actually install the thermal pane units, they are placed on quarter-inch neoprene shims that look like dominos. You get them where the units are made. By positioning the glass equidistant right and left, it is easy to establish a quarter-inch air space all around the unit, standard practice for safety. Finish both sides of the unit with 1-by-1-inch wooden trim or — more expensive — molding made for the purpose.

Opening windows. Purchased or recycled opening windows come in their own lightweight frames, generally three-quarter-inch finished stock, which is not strong enough for laying cordwood masonry against. Once again, make a window buck of dimensions a half-inch greater than the unit itself: up and down, right and left. Later, you will install the window with the aid of tapered wooden shims (like wooden shingles) to tighten the window's own lightweight frame to the buck. Then trim the unit around its edges, both sides. Small packages of these shims are sold at building supply stores.

Cleaning the Log-ends

Mortar does not adhere well to wood, so it is very easy to clean off any excess mortar that might have stuck to the ends or side edges — the "reveal." Wait until after the mortar has set hard, say a week, or even after all the cordwood walls are completed. We have used both stiff plastic-bristled brushes and wire brushes for the purpose. With bottle-ends though — discussed in Chapter 9 — mortar must be cleaned off the same day the bottle units are installed.

Finishing Up for the Day

1. Pick up any mortar which has fallen onto the concrete foundation. Sweep the area. It is so much nicer starting next time at a tidy work site.
2. Cover the top of your work for the night, not the sides. Use plastic weighted down with boards — or whatever else will shed a driving rain from the top — thus protecting the inner insulation cavity.
3. If using wet sawdust in the mortar, be sure to prepare enough for the next day's work. Pass it through a quarter-inch screen into the soaking vessel and add enough water so that the sawdust is totally saturated. You can prepare sawdust 2 to 4 weeks in advance, if you like. No harm. After a month or two, though, it might take on a bit of an odor.
4. Make sure your sand and insulation sawdust is covered with a tarp.
5. And, finally: *Please* don't forget to wash your tools, wheelbarrow, and gloves. And rinse your hands, too, in *clean* water, not water contaminated with cement.

Chapter 9

Finishes and Special Features

One of the "5-E" advantages of cordwood masonry is that it is *Esthetically Pleasing.* Owner-builders have fun creating special design features, whether they are made from log-ends or from their flashy cousins, bottle-ends. We'll cover both of these in this chapter, as well as finishing options for the cordwood wall.

Cordwood Designs

Like stone masonry, cordwood is an excellent tableau for incorporating designs. Specialty log-ends, patterns, shelves ... your imagination (and common sense) are the limits. Large log-ends generally need to be planned for. The chances of the right space appearing exactly where you want it are slim to nil. The most extreme examples I can think of are the six very large (52-inch diameter) *Balm of Gilead* (a cottonwood) log-ends in our La Casita guesthouse.

Specific log-end design features need even more planning. For the flower design of Figure 9.2, I placed a piece of plywood (of the same panel size) on two saw benches and moved vertically standing log-ends around on the plywood until the design looked right, always keeping a one-inch space between elements. Then, I transferred the log-ends from the plywood to the panel, this time with mortar.

For the diamond design of Figure 9.4, I made use of a number of 4-by-4 inch square cypress samples given to me by a timber frame supplier. I drew the design at full scale on a large piece of corrugated boxboard and tacked it to the outside of

9.1: *The large log-ends constitute half the area of these panels. The other 60-odd log-ends — and the mortar — make up the rest. We rolled the 300-pound pieces into place, supported them with wedge-shaped log-ends, tied them off to the frame with temporary 1-by braces, and built the rest of the panel around them.*

the building's frame. (This method was also used for the Big Dipper bottle design in Figure 9.13.)

Figure 9.3 features five log-ends cut from pieces of log cabin logs being thrown out at a local sawmill. In this case, a fairly flat cradle of the right size was created, and then the design built itself. Surround a design with random rubble logs for accent.

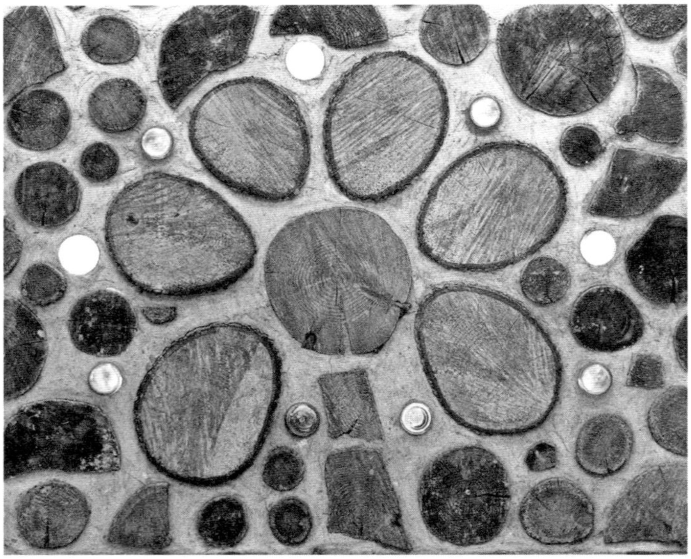

9.2: *The "rough draft" of this flower was done on a horizontal piece of plywood, then transposed to the wall.*

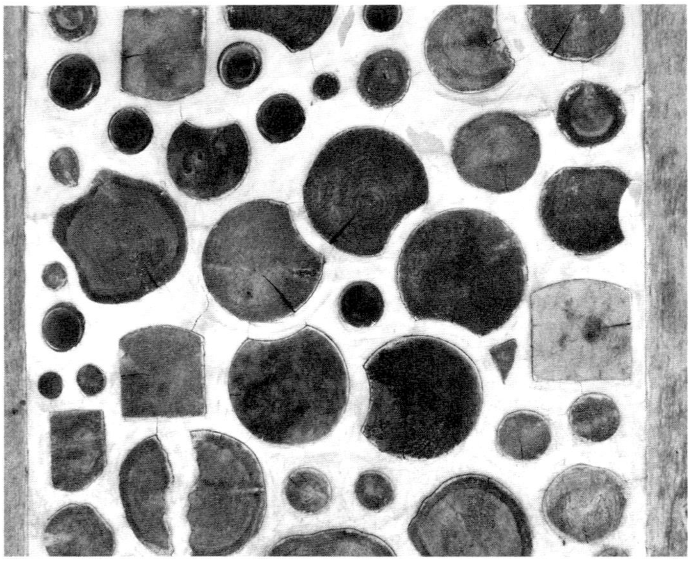

9.3: *A pentagon is made from five equally-sized log cabin logs. The two small round log-ends near the bottom established the right cradle space for the design.*

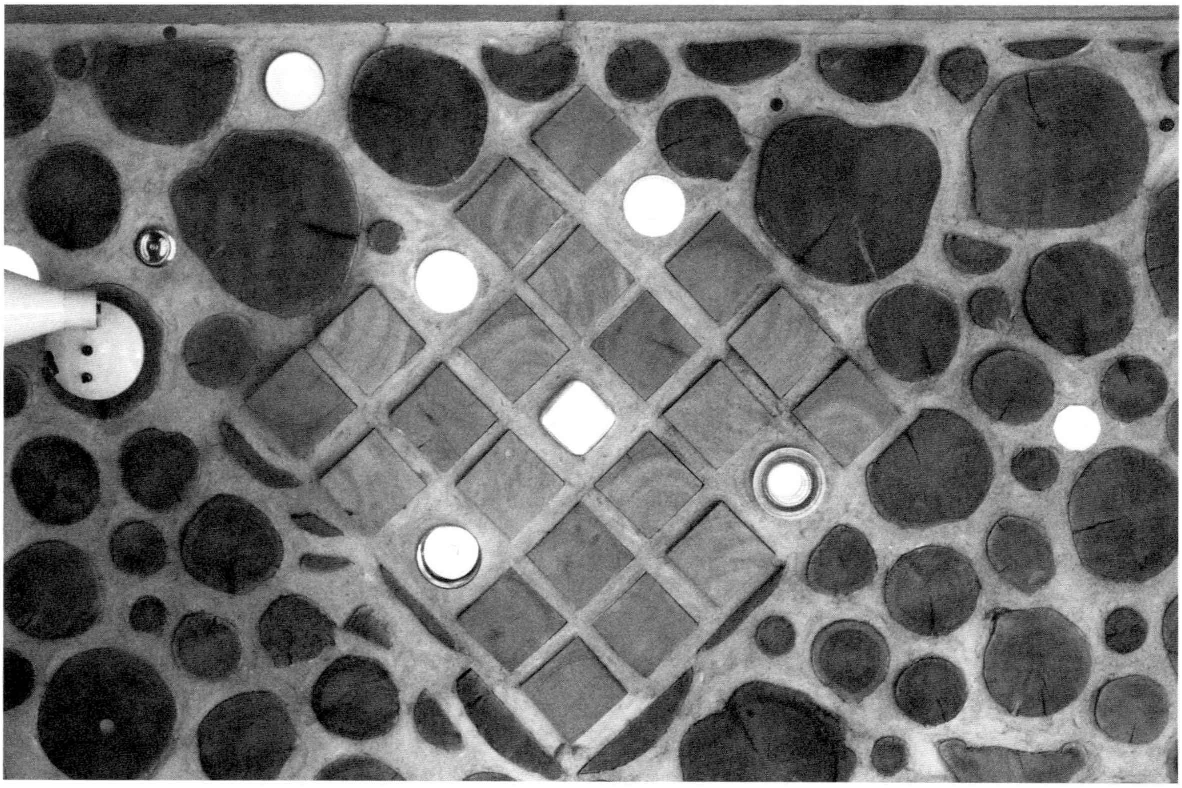

9.4: *A cardboard pattern hanging from the outside of the girt made this design go smoothly. The little "slab-ends" at the bottom helped establish the needed right angle.*

9.5: *This small display shelf is a 24-inch-long half-round log-end extending 8 inches out from the 16-inch cordwood wall. Above the shelf, an inch-thick replica of a 14th-century ivory carving is backed by a shortened (15-inch) log-end of the same diameter.*

Shelves

You can make shelf support pieces from hemispherical log-ends (Figure 9.5), or from round log-ends that have had a piece removed, as per Figure 9.6. In these examples, the support pieces stand alone for a specific item: no long shelf. Figure 9.7 shows support pieces at each end of a shelf made from a varnished 2×8.

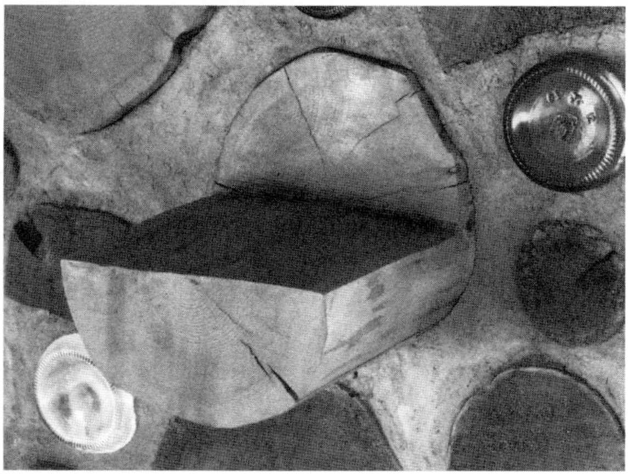

9.6: *Use a log-end 6 inches or so longer than the wall thickness. Cut it half way through with a chainsaw and split off a piece with a hammer and chisel. Sand it up with a circular sander like the one in Figure 9.14. I put two or three coats of water-based urethane on shelf pieces.*

9.7: *This 5-foot long shelf is a varnished 2×8, and supported by two support pieces of the kind seen in Figure 9.6. Use a 4-foot level to place them so that the shelf lies flat.*

Cubbyhole

When cut into logs, cedar and other woods growing in wetlands often have a hollow core for the first few feet of their trunk. These are not much use to a sawyer, but cordwood builders can put bottle-ends in them ... or, make a cubbyhole. We have two of them in our sunroom wall. I cut two hollow log-ends to 8 inches in length. With a wire brush, we cleaned up the deterioration that made them hollow, and placed them up to regular (solid) 8-inch log-ends of the same diameter. We installed this fabricated 16-inch log-end with a cubbyhole to the inside and the solid end to the exterior.

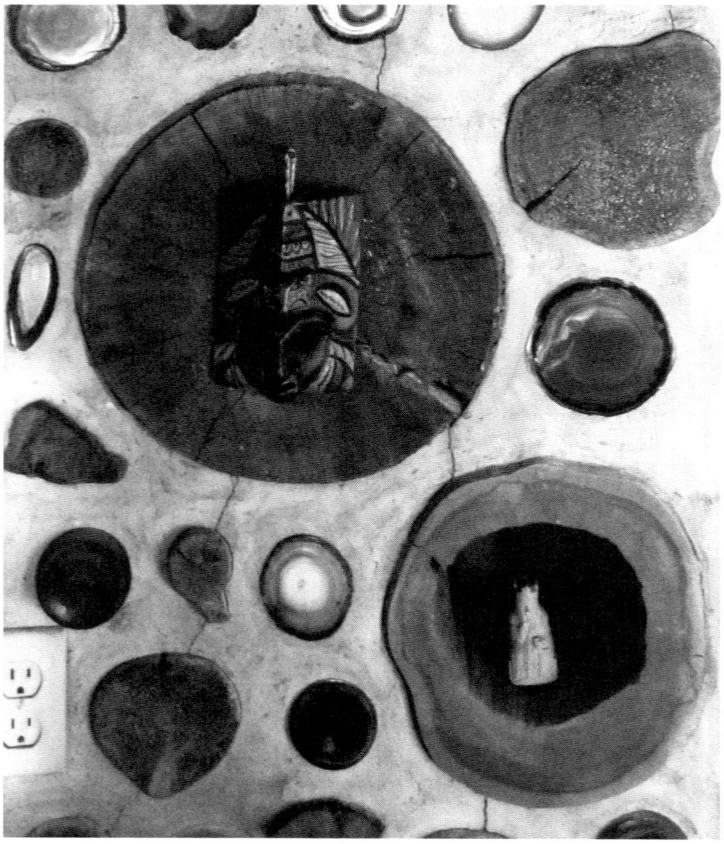

9.8: *A cubbyhole provides a recessed space to display a special keepsake.*

Bottle-ends

Bottles are used with various sustainable building methods such as cob, Earthships, and even straw bale, and they have been used with cordwood masonry since the 1970s. "Cordwood Jack" Henstridge put wine bottles in his 10-inch cordwood walls in New Brunswick in 1974, but he left the necks showing on the exterior. One Halloween, some local yahoos broke the necks off. At Log End Cottage in 1975, Jaki and I began to make "bottle-ends" (some call them "bottle-logs") from two cylinders cut from bottles. No necks to break!

Bottle-end designs are like log-end designs ... but much more colorful! We call them poor man's stained glass. People seem to like them, so you'll find bottle-ends in most cordwood buildings. We've never done one without them.

Making bottle-ends the easy way

It is important to know how to make bottle-ends so that they will work well with cordwood masonry and last indefinitely. There are two distinct methods for making high-quality bottle-ends.

Method One: Cut and Tape

With wall thicknesses of 8 inches to 12 inches, bottle-ends are made from two open cylinders — "tumblers" — of the same diameter that are fastened together with duct tape. Usually, this involves cutting the necks off the bottles to reduce them down to a regularly-sized cylindrical tumbler. But how? In the early days, we tried various folk methods, all with poor success. But for over 20 years we have had a better than 95 percent success ratio cutting bottles with a slate and tile cutter, which has a diamond-tooth circular saw that rotates through water to keep it lubricated. Buy one on sale for $50 to $60. And you can use it for cutting tiles, too, its intended purpose.

Because almost all bottles have a quarter-inch chamfered bottom surface, we tend to make our bottle-ends about a half-inch longer than log-ends, and leave a quarter-inch sticking out on each side of the wall. This facilitates pointing around them and gathers more light. Simply set the saw's stop (the plastic bar on the left in Figure 9.9) to a quarter inch greater than half the width of the wall: 4.25 inches, for example, for walls of 8-inch thickness. When joined, the tumblers will form an 8.5-inch bottle-end.

After cutting bottles with the tile saw, soak the tumblers in a bucket with a mixture of water and five percent bleach. Use a long-handled scrub brush to clean away any yeasty dregs still clinging to the bottom of the tumbler. Rinse the tumblers in clean water and turn them upside down to drain for five minutes. Then turn them right side up and put them in the sun. When fully dry, place them together and duct-tape them around the middle.

9.9: *Slate and tile saw used for cutting bottle-ends.*

9.10: *Bottles have been cut into tumblers (except for two jars, useful as they are), and are drying in the sun. Keep your pairs together throughout the process. Remove labels if they come off easily during the washing process. At the very least, remove any label material within an inch of the end of the bottle, so that it does not show. A razor blade knife works well for this.*

Back in the 70s, we'd glue two clean and dried tumblers together, but we found out that when the unit is sealed *too* well, there is a risk of air inside expanding, which can break the bottle. Now we make sure there is a way for air to escape. After taping the two tumblers together, we puncture two or three holes in the tape so that air can escape into the insulation cavity.

If you happen to have a clear jar of the same diameter as a colored bottle you want to use, simply adjust the stop on your saw. If you have a 5-inch tall jar, you would cut the companion bottle at 3.5 inches, giving the desired 8.5-inch bottle-end.

Method Two: Plug into a Cylinder

If the wall thickness is 16 inches or greater, plug two uncut bottles — or a bottle and a jar — into a flexible cylinder made from an offset printing plate, aluminum flashing, old vinyl or some similar strong, flexible material. (With whole bottles plugged into a flexible cylinder, soak dirty bottles in the bleach solution, then shake them up to get rid of mold, which can grow inside if not removed. Drain and dry the bottles, which might take 24 hours.)

Recycled aluminum printing plates are our favorite; they're free — or very cheap from small print shops — and have just the right flex. (With stiffer aluminum flashing, taping is usually required.) The basic technique is the same with any of the flat materials. Cut them to a rectangle which, when rolled into a cylinder, is long enough to grab the cylindrical part of the bottle by two inches and wide enough that when it is wrapped around the two bottles, it will lap onto itself by an inch.

Use a couple of strong elastic bands to create a "spring-loaded" cylindrical bottle-end holder, which holds the bottles firm until they are put in the wall. If they are loose in the cylinder, use a little duct tape — printing plate to glass — to assist the elastic bands. As the cylinder is not airtight, we don't have to worry about excess steam pressure. Leave enough glass exposed, at least three inches, for the mortar to bond to the glass at each end.

Bottle-ends are laid up in the wall the same as log-ends, but there are some tips we can pass on from years of experience.

- Combine a clear tumbler with a colored one, to maximize light transfer. A bottle-end made from two dark bottles transfers much less light. So, about half the bottles you collect should be clear. People are excited about collecting pretty blue and green bottles and find themselves short of clears to match them with. An exception to the clear-and-color rule would be two light-colored tumblers, such as from two yellow wine bottles.
- The color is more vibrant when viewed through the colored end; it looks diffused if the bottle-end is turned the other way. As we usually want to enjoy the design feature from the building's interior, this means colored ends are laid to the wall's inner surface. An entranceway might be an exception, where you want to greet night visitors with an illuminated light design.
- With Method Two, place the bottle-end with the cylinder's overlap down, stopping

9.11: *A green and a clear bottle of the same diameter can be plugged into a flexible cylinder and held fast with two strong elastic bands.*

9.12: *Adjusting the length of the cylinder allows for bottle-ends from 16 to 24 inches in length.*

insulation in the wall's cavity from finding its way in.

- Bottle-ends take time and care to do right. Assembly-line the process and get them all made ahead of time. Don't slow down wall building to make bottle-ends. Pair the clear bottles off with colored ones of the same diameter *and keep them in pairs throughout the process,* as seen in Figure 9.10.
- It is important to clean the ends of the bottles the same day you lay them up! Mortar bonds with glass, but you can remove mud that smears onto the bottle-end during building or pointing. Use a clean dry cloth. For the little raised dots found at the bottom of many bottles, an old toothbrush works well. After the first day, you will need a ten percent muriatic acid solution to remove dried, smeared mortar. It works, but it's best to avoid having to do that.

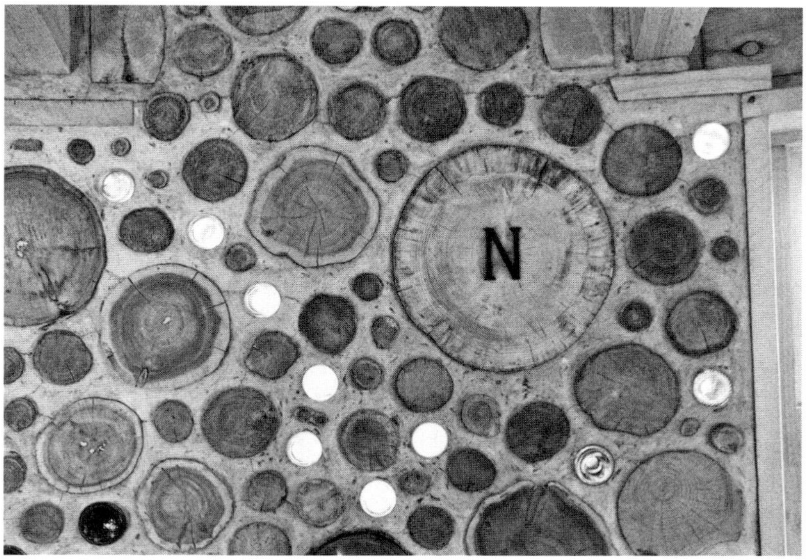

9.13: *The constellation — and the North Star — were drawn on a large piece of corrugated boxboard which was hung from the radial rafters of our round Mushwood Cottage. The stars are all made of clear bottle-ends in this 12-inch wall. The two pointer-stars point to Polaris, the North Star, upper right, which marks true north. The large 17-inch diameter pine log-end with the big N is nominal north. Three other, similar log-ends mark E, S, and W.*

Bottle-end designs

Our first bottle-end "designs" were constellations. Whenever a space appears that is the right size for a bottle-end, you can put it in ... or not. Stand back from the wall once in a while to see if you are creating a pleasing balance of colors and placement. Sometimes two bottles next to each other looks right. The end result is your own constellation, unlike any in the sky.

Or, you can re-create a genuine constellation, such as Ursa Major — the Big Dipper.

Other bottle-end designs are displayed better in the color section. For some advanced techniques, see *Cordwood Building: A Comprehensive Guide to the State of the Art* (New Society, 2016, pages 91–98). For remarkable owner-designed examples from around the world, Google "cordwood masonry bottle designs."

Cordwood Finishes

In the vast majority of cordwood buildings, no finish is applied to the cordwood masonry. The cleaner, brighter end of the log is placed to the interior, where it will look pretty much the same as the day it was laid up for 30 or more years. (For this reason, it is good to use a freshly-cut end for the interior.) The weathered, darker, or less attractive end is placed to the exterior, where, in three years or less, it will take on a rustic weathered appearance: gray, red, black, or brown, depending on wood species and the compass orientation of the panel. This weathering is not deterioration, just a change in color. (Out of all the cases I know, only one cordwood builder, Sandy Clidaras at his "Cordstead" just north of Montreal, has successfully "preserved" a finished appearance of the log-ends on the exterior. Sandy's method involves several time-consuming steps, including dipping the log-ends in borax and "environmentally safe" propylene glycol. If you want to go that route, go to thecordstead.blogspot.com and order his excellent CD, "The Cordstead Collection," for full details and pictures.)

The various "finish" options listed below are the exceptions to common practice, not the rule. Jaki and I have used some of them ourselves.

9.14: *The author sands log-ends with his favorite power tool, a Makita 4,500-rpm circular sander.*

Sanding and finishing log-ends

Sanding can serve two purposes. On the interior, it prepares the log-ends for the application of a finish. On the new second story of our Mushwood summer cottage, I pre-sanded most of the larger (8 inch to 10 inch diameter) log-ends and pre-applied two or three coats of semi-gloss Minwax Polycrylic Protective Finish, a water-based urethane. The wall looks great: bright and clean with a nice cedar color.

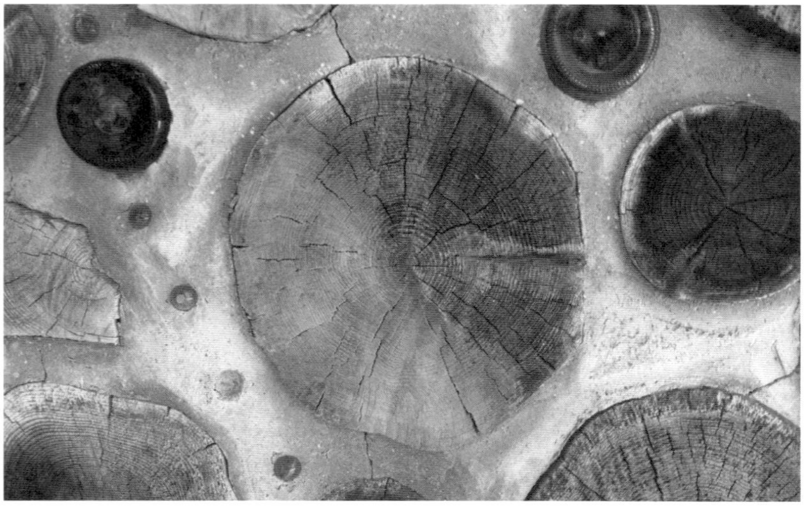

9.15: *This log-end is half sanded, to show contrast.*

On the exterior, I'll sometimes sand old, weathered log-ends to brighten the wall. The sanded log-ends will re-weather with time, but never as much as the first time. The sanding seems to reduce weathering. But if you really want to make an impact on the exterior, apply a coat of Cabot Waterproofing, "an advanced technology silicon sealer." It is much nicer to use and will last six times longer than regular chemical or petro-based waterseals. Several years ago, I sanded the entire north side of our little Hermit's Hut guesthouse — out in the thick woods — and applied Cabot Waterproofing on the log-ends as well as to the 8-by-8 inch corner posts. It still looks great and protects the wall from moisture. The product is for exterior use only. And do not use the product without sanding first with #36 or #50 grit paper. Unsanded log-ends will suck up the sealer like straws.

As discussed back in Chapter 5, log-ends last because they breathe on end grain. Conditions for fungal growth are not present. If one end of a log is sealed, it can still breathe the other way. This is why I am not a believer in sealing both ends of a log, particularly if sap moisture is still present. In completely seasoned wood — with only ambient moisture content — it *might* be okay to seal both ends, but I have never done this.

Far and away the best tool for sanding log-ends is a high-speed disk sander. I use a Makita 4,500-rpm 5″ disk sander, seen in Figure 9.14. Be sure to wear eye protection and use the safety handle attached. Be careful. I have used this tool a lot over 40 years, and love it … but have cut myself more than once.

Here are some sanding tips: (1) The easiest time to sand log-ends is when they are being held firm in a finished wall. Recessed pointing keeps the mortar out of the way of the sanding disk. (2) If you wait a year or two to sand, you can clean up weathering. A silicone sealer does a very good job of keeping the log-ends looking good for a long time. With quaking aspen — called "popple" in New York's North Country — the blackened ends, caused by bacteria digesting wood sugars near the surface, can be eliminated and, as the sugars have now been digested, it doesn't come back.

Sheetrocking over cordwood

Back in the 1970s, I went to one of Jack Henstridge's talks on cordwood masonry. A lady asked, "Mr. Henstridge, can I put lath and sheetrock over the cordwood walls?" Without hesitation, Jack came back with, "Yes, ma'am, you certainly can. And then, to give the wall that authentic appearance, you could put on cordwood wallpaper." I can't improve on Jack's answer.

Painting cordwood

What? Well, it has been done, and quite nicely, too. One family painted all of their internal cordwood walls with light-colored latex paint, two coats. Some rooms were white, some very light blue. The relief of the masonry was still there, but the rooms were infinitely brighter than with the original very dark poplar (quaking aspen) cordwood walls.

To brighten a dark bedroom, Jaki painted part of the cordwood walls with a 50/50 mix of ivory latex paint and white joint compound, as seen in Figure 9.16. She caulked any large gaps and applied two coats for good coverage. Some of the wall was left unpainted to demonstrate the effect.

Siding over cordwood

Cordwood homes built in the 19th and early 20th centuries were sometimes sided after the house was built, possibly a few years later. Jack Henstridge used to say that people covered up

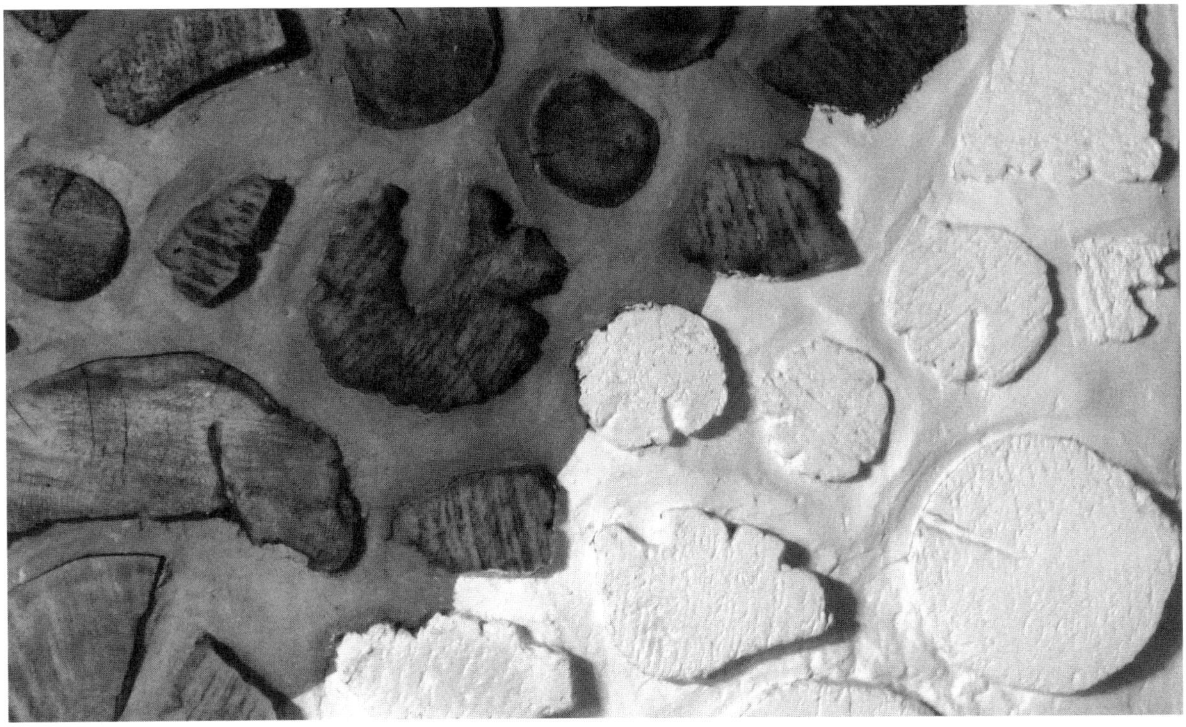

9.16: *The texture and relief of the cordwood wall is still evident where the wall is painted.*

their cordwood walls to try to hide that they were living in "poor people's housing." He told me that during the 1920s cordwood masonry was the prime method used to rebuild the town of Saint-Quentin, New Brunswick, after a disastrous fire. Many of these buildings were later covered with wood siding. When Jaki and I traveled to Winchester, Ontario, in 1975 to research cordwood masonry, we found cordwood houses that had been sided. Since cordwood's strong rebirth in the mid-70s, however, I know of no instance where cordwood walls have been covered with siding. One of the main reasons that people choose cordwood is that they like the look of it, and people are more careful now about wood selection and drying, so there have been fewer problems that might prompt someone to cover the external walls. That we have modern ways of dealing with any wood shrinkage that might occur — covered in this book — may be a factor as well. I am always reluctant to suggest how to do something that I have not done myself — or even seen done in modern times — but it strikes me that it would be a fairly easy matter to fur out a cordwood wall and side over the furring with insect screens at the top and bottom. I don't see any breathability issues arising.

Plastering over cordwood

Jaki and I have seen two successful plastered interior cordwood walls in person, one in Washington and one in Ontario. These were done around 20 years ago but, unfortunately, the owner-builders can't remember the details of how they did it. However, more recently, our friends Peter and Blythe in Tasmania plastered their interior walls — they say "rendered" — and they're very pleased with the results. They tell the story:

Rendering Cordwood in Tasmania

by Blythe Tait and Peter Robey

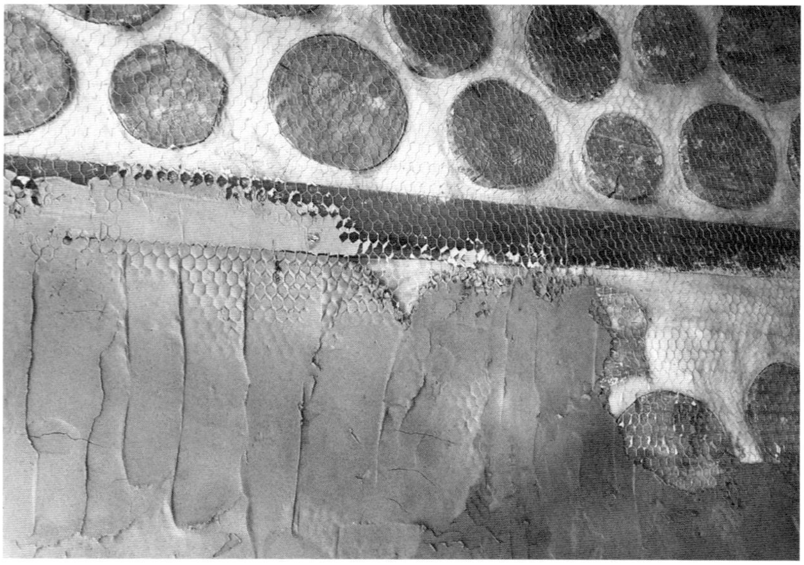

9.17: *The first coat covered the log-ends and the large recessed spaces in the mortar between log-ends.* Credit: Peter Robey.

We lived in a 20-foot square cordwood cabin for five years, with walls only eight inches thick. We had used green cordwood, so after the log-ends dried out, we were left with massive gaps around them, large enough to see through! Indoors, the logs provided perfect little shelves for collecting dust. Finally, while we love the appearance of cordwood on the outside of a structure, we both prefer a less busy interior, and this is what we decided to do on our main house, a two-story 16-sided building. Here's how:

Once we had finished all the cordwood masonry walls in our main house, we rolled out small-gauge galvanized chicken wire onto the walls, stapling it in place onto the log-ends. Then, we mixed up a blend of sand and lime in a ratio of 3:1 in a cement mixer, adding enough water to give a spreadable consistency. We let each batch mix for at least 20 minutes and then let it sit another 20 minutes before using it. This render — the Aussie word for plaster — was sticky and couldn't be poured out of the mixer. We used our hands to slide it out of the mixer and applied it onto the wire-covered walls with trowels. This first layer was very thick in some places. It covered the logs completely in a thin layer and covered the spaces in between the logs with render up to 1.5 inches thick. Once dried, we could still see the impressions of the chicken wire.

Once the first layer had dried, we applied a second layer. We changed the ratio of sand to lime to 1:1, hoping for a finer, smoother finish, but this blend simply didn't work for us. It didn't adhere well to the first layer of render; we found that a few firm taps on it would bring sheets of the render crashing down! It was easily removed with a shovel, leaving the first layer still intact and unaffected underneath.

We went back to our original mix, which adhered well to the first layer. This meant that our render remained very coarse, with lots of "character," but we loved the look, so it wasn't a problem. Anyone wanting completely smooth walls will need to do much more experimenting with render than we did.

Once the render was dry, we mixed some lime and water to the consistency of full cream milk and applied this limewash in layers using paintbrushes. The limewash filled any little cracks in the render, basically giving the walls a shell, and we feel as if it provided another layer of strength to the walls.

Voila! Three years later, we are still thrilled with the effect.

9.18: *The cordwood masonry is covered with plaster rendering.* Credit: Peter Robey.

(*Author's note 1:* Jaki and I conducted a cordwood workshop at Peter and Blythe's place in 2010. Much more about their project appears in *Cordwood Building* [New Society, 2016], Chapter 20: Hexadecagons in Hawaii and Tasmania.)

(*Author's note 2.* It is my view that a lime plaster greatly reduces air infiltration, but is still "breathable." I would not use Portland cement in this application. Further, while plaster was applied to exterior walls in historic cordwood buildings, it invariably fell off of the wall over time. I would restrict its use to interior applications, as per Peter and Blythe.)

Chapter 10

Maintenance and Renovations

Cordwood masonry — done correctly — requires very little maintenance. Weathering is not deterioration, and is not really a maintenance problem. It is simply part of the rustic cordwood style. Nevertheless, I have had good success with sanding exterior log-ends with a circular sander, and applying a siliconized sealer, as described in Chapter 9.

Log-end Shrinkage Solutions

A common problem is log-end shrinkage a year or two after the masonry is completed. In fact, waiting a couple of years before attending to shrinkage may very well prevent having to deal with it more than once.

Primary check expansion. We use mostly round white cedar log-ends nowadays, but, even

10.1: *A large primary check needs to be filled.*

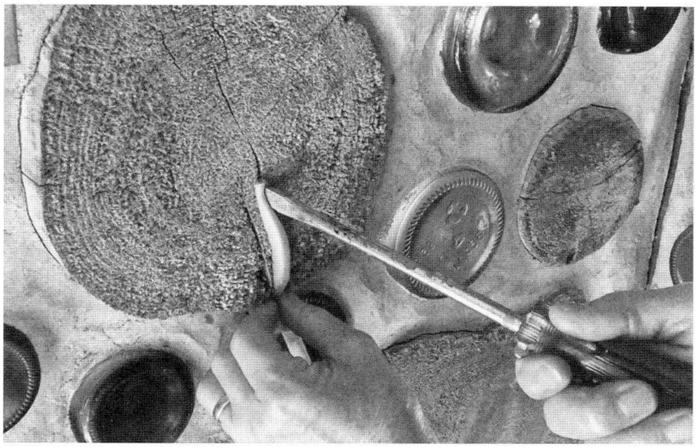

10.2: *Push backer rod into the check with a flathead screwdriver.*

10.3: *Caulk with clear siliconized caulking.*

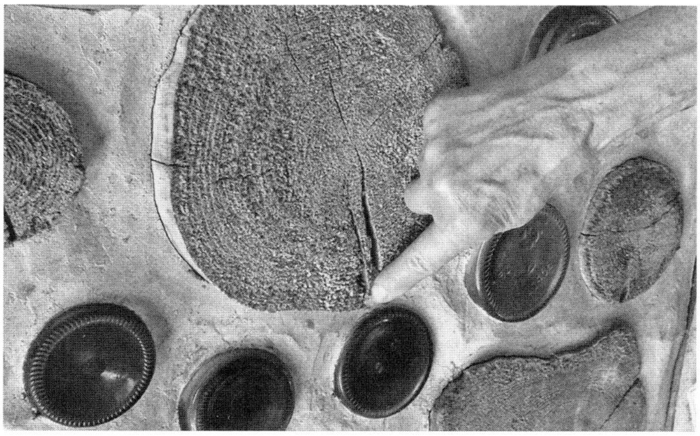

10.4: *Jaki seals the caulk with her finger. The end result looks very much like a round log with a check in it. In fact, it looks exactly like Figure 10.1.*

though they have about the lowest shrinkage rate of available woods, the primary check in the larger ones — 6-inch diameter and greater — can open up enough in the year or two after construction that you can see daylight through it. Before caulking the check, we stuff it with gray *backer rod,* a 25-foot coil of foam that is used to close off drafts around doors and windows. It comes in various diameters, but the ⅜-inch size works well with cordwood. With a flat-headed screwdriver, we stuff the backer rod into the primary check, leaving it recessed about a half-inch. Then we apply a bead of clear caulking over the backer rod, almost level with the log-end surface. I like clear caulking for this purpose — your chances of success in matching the color of the mortar are slim to nil. A check looks natural in a round log, and, with clear caulking, the check still looks like a check. A siliconized caulking, such as Red Devil Lifetime caulking, is less expensive and less messy than pure silicone caulk.

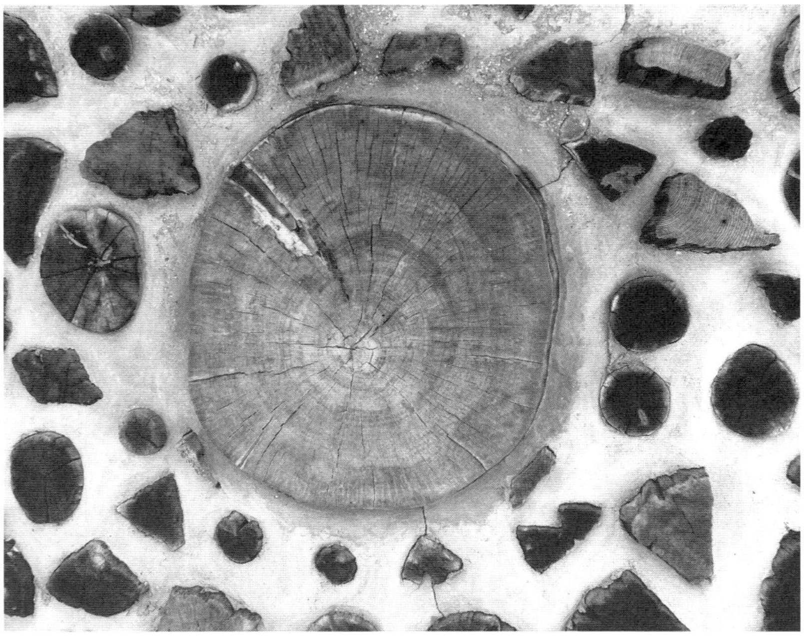

10.5: *This 16-inch beech log-end shrunk a good quarter inch after a few years, but was repaired as described in the text.*

Radial shrinkage. The larger the log-end, the greater the width of radial shrinkage. Because Jaki and I have always used well-seasoned log-ends, the only ones that we've had to attend to for this kind of shrinkage have been the really big ones used as design features, particularly the hardwoods. At Earthwood, we have six large (11-inch diameter) elm rounds and, next to our front door, a lovely 16-inch beech. Radial shrinkage grew to about a quarter inch around the edges of all of these, even though the beech log-end had been drying for three years near a woodstove prior to use. The fix is essentially the same as with the primary check expansion described above: stuff in some backer rod, and then caulk the gap with clear siliconized caulk.

Mortar Cracks

There are fewer mortar crack problems in cordwood walls nowadays, in large part due to the mortar recipe advice that we and other experienced builders have given to newcomers over the years. Mortar shrinkage cracks still occur, however, due to:

1. **Lack of mortar retardant:** The addition of the appropriate soaked softwood sawdust or a commercial cement retarder slows the set, which greatly reduces mortar cracks.
2. **Direct sun:** Sun beating down on the work can dry even good mortar too quickly. Always try to work on the shady side of the building, or build a protective shelter from the sun.
3. **Over-pointing:** This pulls water to the surface, causing rapid drying, which results in surface cracking. The problem is further exacerbated by pointing too soon.
4. **Mortar mixed too wet:** The more water that transpires from the mix, the greater will be the shrinkage and cracking.

Even if you do everything wrong — both the wood and the mortar suffer severe shrinkage — this is not a structural disaster, and you can still fix it.

Consider: Despite mortar cracking and wood shrinking, the thousands of masonry units — log-ends and mortar sections — fit together more perfectly than any dry-stone wall this side of Machu Picchu. If built within a timber frame, the walls are in no danger of tumbling down. But it doesn't look very nice, and there will be more air infiltration.

The solution is the application of one of the products made for log cabin chinking. One brand I have experience with is Log Jam Chinking from Sashco. Other cordwood builders have used Perma-Chink's log cabin chinking ("Log Home Sealant,") and Weatherall's Triple-Stretch chinking ("Textured Log Home Sealant"). Applied to log cabin chinking or to the space between log-ends on a cordwood wall, these products will fill gaps of a quarter-inch and span mortar cracks. The products stay flexible and will move with any further expansion and contraction of the wood. Buy it in five-gallon pails — much cheaper than in caulking tubes. These are acrylic latex products.

You can apply these chinking products to the interior or exterior mortar joints, or both. Appearance and cost are the main considerations. The cost is between $210 and $260 for a five-gallon pail.

The possible need to apply some sort of repair, due to mortar cracks or log-end shrinkage, is yet another good reason to leave the logs "proud" of (protruding from) the mortar matrix by a good quarter-inch. This "reveal" makes application of chinking products much easier, with much less likelihood of smearing it on the ends of the logs. To apply the product, take the cover off the pail and dip your pointing knife into the gooey chinking. Pull up a comfortable working amount and spread it on the mortar joint, pushing it with the knife into any radial shrinkage gaps. You can greatly extend the coverage by drawing the material along the mortar joint with a small paint brush, dampened with water. A sixteenth of an inch thickness is enough to bridge mortar cracks, and this thin layer is easier to feather out with a brush than to create with the pointing knife alone.

Coverage will depend on the width of the mortar joints, the severity of wood shrinkage, and your skill with application. Assuming a ¹⁄₁₆-inch application thickness and walls that are 60 percent wood and 40 percent mortar, you should be able to repair 500 to 600 square feet of wall area with a five-gallon bucket of Log Jam. Log Jam is smooth, has excellent adhesion, and is very flexible even after ten years on an exterior application.

One of the extra benefits of using a chinking product in this way is that color irregularities from different mortar batches are returned to a single consistent color. And, if you choose a light color, like Log Jam's "white white," the cordwood wall will be very much brighter and light reflective. For this reason, you might consider starting with the interior first, and learn what your actual coverage is with a five-gallon pail. Just doing one side of the wall will greatly reduce air infiltration around shrunken log-ends. Then you can decide about doing the exterior, with a view to improving appearance, getting even more infiltration protection, and discouraging insect attack.

Another effective way to greatly reduce air infiltration in a cordwood wall is to plaster — or "render" — its interior surface, as described in Chapter 9.

Chapter 11

Building Codes and Permits

MOST CORDWOOD HOUSES have been built by owner-builders on rural land without a building permit. We built Earthwood in 1981–82 before our town even had building permits. We did get a permit for the Stoneview Guesthouse and for our Mushwood Cottage on Chateaugay Lake. But what if *you* need to get a permit?

The International Residential Code (IRC) is in use or has been adopted in 49 states (excepting Wisconsin, see below), the District of Columbia, Guam, Puerto Rico, and the U.S. Virgin Islands. While cordwood masonry is not specifically mentioned in the IRC, neither is it proscribed. Rather, cordwood, like some other natural building methods, falls under Paragraph R104.11, where it says, in part: "The provisions of this code are not intended to prevent the installation of any material or to prohibit any design or method of construction not specifically prescribed by this code, provided that any such alternative has been *approved.* An alternative material, design or method of construction shall be *approved* where the *building official* finds that the proposed design is satisfactory and complies with the intent of the provisions of this code, and that the material, method or work offered is, for the purpose intended, not less than the equivalent of that prescribed in this code."

In short, the builder needs to convince the building official that the method proposed meets or exceeds the requirements of the code.

Wisconsin, with its own Uniform Dwelling Code (UDC), is a hotbed of cordwood construction. Richard Flatau, of Cordwood Construction Resources, LLC in Merrill, Wisconsin, told me about the Cordwood Education Center, an architecturally drawn, community-built, privately funded cordwood shelter for a local school system not far from Richard's own cordwood home. Jaki and I have visited the center, which exudes quality in every way. Richard tells me: "The code process was attached to a certified architectural firm. There was no rush. As it turned out, the conversation I had with the head of the Wisconsin UDC helped tremendously. With this state-approved public school classroom under our belt, it has given cordwood more legitimacy in the code and building industry and has become the bellwether to introduce code officials to the cordwood process." Further, Richard says: "Wisconsin requires that all homes meet code requirements, even if there isn't a code official nearby. A family home *has* to be code compliant or they can turn your power off, withhold an occupancy permit, fine you, or make you take it down."

Bruce Kilgore, Alan Stankevitz, and many others have secured building permits for their cordwood homes. Bruce went to his town's code enforcement officer, and "asked what he wanted to see, his interpretation [of how the code might apply], and what he wanted us to do" to get a permit. The CEO took an interest in the project and made several visits to the site during construction. "Basically," says Bruce, "he was concerned about the usual health and safety issues. Simple stuff." Bruce's brother-in-law did the engineering on the timber framing to show it would exceed load requirements, and Bruce provided documentation that the cordwood masonry infill would greatly exceed New York's Energy Code requirements, which is R-19 for walls.

11.1: *The Cordwood Education Center, near Merrill, Wisconsin.* Credit: Richard Flatau, Cordwood Construction Resources, LLC.

Of specific benefit to the owner-builder seeking code approval for a cordwood home are two documents: *Cordwood and the Code: A Building Permit Guide,* (2005, Richard Flatau and Alan Stankevitz, with Rob Roy and Dr. Kris Dick), available from Earthwood Building School. This 56-page document, a product of the Continental Cordwood Conference of 2005 in Merrill, Wisconsin (CoCoCo/05), has helped numerous cordwood builders through the permitting process. Sections include: Richard Flatau's Conversation with a Wisconsin Code Official; Thermal Monitoring and Fire Resistance of a Cordwood Wall, Certified Compression Tests of Cordwood Mortar; Cordwood in a Seismic Three Zone; REScheck Energy Code Compliance; and more. In the Introduction, the authors say: "This guide contains various reports and tests which should put the code official's mind at ease regarding the integrity of a cordwood dwelling. How you present this to a code official is entirely up to you. Try not to overwhelm them with too much information or underwhelm them with not enough. With that in mind, the [17-page] Appendix A is a sample of what you might submit to a code official. This document is yours to modify to your heart's content."

In point of fact, Appendix A is a copy of the successful building permit submitted by Alan Stankevitz to his own code official. The CD (included) allows you to adapt it for your own application.

Of particular import is Kris Dick's report on thermal tests that he conducted with sensors placed within a 24-inck thick wall at the University of Manitoba. His Summary/

Conclusion says, in part: "Based on approximately three months of mid-winter temperature data, the wall was determined to have an RSI Value of 6.23 (m^2K/W), R-35 for a 24-inch (60-centimeter) wall system."

I know of several cases where *Cordwood and the Code* has eased the permitting process for cordwood owner-builders.

The other useful document is the 15-page Chapter 24 of my *Cordwood Building: A Comprehensive Guide to the State of the Art,* (New Society, 2016), entitled "Getting a Building Permit for a Cordwood Home" — too long to reprint here, but useful to those builders who might need to go through the permitting process. Sub-sections include "An Engineering Viewpoint" by Dr. Kris Dick, "A Code Enforcement Officer's Viewpoint" by Thomas M. Kwiatkowski, and "Other Cordwood Code Issues" by Rob Roy.

Chapter 12

Tools

Power Tools for Harvesting Cordwood

REALISTICALLY, in today's world, cutting trees into log-ends with an axe or any kind of human-powered saw is no longer a sensible option — given the labor-intensive nature of cordwood masonry in general. Of course, it would have been mandatory from — say — the year 1000 up until sometime in the 19th century.

Chainsaw

Cutting trees in the forest is most easily accomplished with a chainsaw. While it is beyond the scope of this book to provide instruction in the safe operation of chainsaws, two points are critically important.

Firstly, your chainsaw will be getting an awful lot of use, so get a good, reliable one. My experiences with lesser — cheaper — brands tell me that I should have gotten a more reliable saw from day one. I have been very happy with the Stihl line of saws; but of course, Stihl is not the only choice. Ask professional wood cutters in your area what they recommend.

Secondly, get training in the use of the saw from an experienced operator. Best is to take a course in safe chainsaw operation, including how to safely fell a tree. Chainsaws can be dangerous. So can tree-felling.

The chainsaw is used to cut the trees down, limb them, and cut them into a convenient length to carry out of the forest, usually 50-inch or 100-inch lengths, depending on their weight. With 50-inchers, you can get six 8-inch log-ends, or four at 12 inches, or 3 at 16 inches, or two at 24 inches. These are the common log-end lengths. And you'll have enough length to accommodate the chainsaw's kerf and trim the ends. (*Kerf* is the term for the wood lost when a chainsaw cuts through a log, typically one-quarter inch.) The numerical advantages of 100-inch logs are similar. Clearly, you also need a *tape measure.*

The chainsaw is used again for cutting the logs into log-ends, as per Figure 12.1.

12.1: *This 50-inch log is marked into 16-inch log-ends, with the quarter-inch chainsaw kerf clearly showing. The first cut cleans and trims the end. Make sure the log is supported in a sturdy manner and well clear of the ground before cutting.*

Other cutting options

There are two other options for cutting long logs into short log-ends. One is a 24–30-inch diameter buzzsaw, fixed in place, with a hinged table upon which the long log is placed up to a stop fixed to the end of the table. The entire table tilts into the rapidly rotating blade, which easily slices the long log into short log-ends. The buzzsaw is most often powered by a belt fastened to the *power take off* (PTO) of a tractor, although I once hired one that derived its power from a generator. The buzzsaw yields very clean, straight, smooth cuts, all of the same length. This buzzsaw option is not illustrated; they are not as common as they were even 25 years ago, but farmers still use them in rural areas.

The other option is described in *Cordwood Building: A Comprehensive Guide to the State of the Art* (New Society, 2016). Bruce Kilgore and I devote the entire six-page Chapter 15 to the detailed building and use of a table made specifically for cutting long logs into precise log-ends of any size with a chainsaw. It is seen in Figure 12.2. There is quite a bit of work — and skill — in making this table and the mechanical hinged bracket to which the chainsaw is married, but, for large projects, the time and effort will pay huge dividends. If interested, find *Cordwood Building* for complete instructions. I am very fortunate that I have not had to make one for myself. Friend Bruce graciously loans me his!

12.2: *The author places a log up to the adjustable stop and cleanly cuts a 16-inch log-end with the chainsaw, hinged to the table by the use of a pair of "pillow blocks" — ball bearing mechanisms mounted to sturdy flat-bottomed housing. A counterweight makes it possible to use the chainsaw easily with one hand. The table/saw unit is designed for safety: the saw cuts through a slot in the table and can never hit anything to cause it to bounce back in the direction of the operator.*

Tools for Barking the Cordwood

Barking — or *debarking* (same thing) — means to remove or strip the bark off of the long logs prior to cutting them into log-ends. It is a necessary part of the cordwood-building process. If the bark is left on, the wood will eventually shrink away from the bark, leaving a nice gap for wee beasties to take up residence. (But see the caveat under "Rot in Wood," Chapter 5.)

The first thing to know about barking is that it is infinitely easier to remove the bark in the spring when the sap is rising, than in the autumn, when it is falling. The rising sap in northern New York would be from early March (maple sugar season) right though to the end of June. In areas south of us, you might take a month off these parameters. The rising sap is like a greasy layer that separates the bark from the outer layers of the wood itself.

Almost any sharp tool will work for barking the wood when the sap is rising. My favorite is my pointed mason's trowel, which is also handy later during the wall construction. Just lift the bark off at one end of the long log with the point of the trowel — or knife or scraper — and then you can pull wholesale strips off the log, sometimes over its entire length.

A word of warning: If you cut all the trees down — even at the right time of year — and put them in a pile for debarking later, you may be severely disappointed. I advise people to get the bark off within a couple of days of felling the tree, even in the spring. After a couple of weeks, the nice greasy sap layer that made it easy to get the bark off turns into a kind of glue, and then you need a drawknife.

If, by circumstance or poor planning, you missed the best time for barking logs, there is another option. Soak the logs in a pond or stock tank for a few days. Upon their removal from the water, you will generally find the bark to be much more cooperative.

Drawknife

A drawknife is a tool with a blade between two handles. The very sharp blade faces the user, who pulls the tool toward him/herself, cutting the bark off. To make the job easier — and avoid backache — set the log up on a sturdy support, as seen in Figure 12.3.

A drawknife will work, but doesn't do so well when little branches join the main trunk. Trim these "nubs" right close to the long log for ease of drawknifing. You want them off anyway, as they get in the way when it comes to actually mortaring the log-ends into place. See Figure 12.4.

Tools for Log-end Preparation

Handsaw

A sharp handsaw is handy for cutting off small nubs.

12.3: *Steve Coley removes bark with a drawknife. The tops of the posts are notched, and the rear post has a stop carved into it to provide resistance, thus holding the log steady.* Credit: Barbara Coley.

Scraper

A chainsaw often leaves little quarter-inch log "hairs" on one side of the cut. These get in the way of pointing the cordwood wall later on. They are easily scraped off using a Stanley Surform Shaver — or equal generic brand — as seen in Figure 12.5.

12.4: *Use a chainsaw — or a handsaw — to trim the "nubs" right close to the log.*

12.5: *Holding the rasping surface of the scraper at a 45° angle to the log-end, cut the little quarter-inch "hairs" off the edge with a firm pulling motion.*

Tools for Mixing Mortar

Rubber Gloves and Facemask

First and foremost! Do not handle cement or lime with bare hands, either at the mixing stage or when laying up the mortar in the cordwood walls. These materials are highly alkaline and can cause nasty little *cement burns* in the skin, which become painful and take forever to heal. They can actually develop into a serious medical condition. For an antidote, we have had good results with an early application of 20 to 30 percent white vinegar in water solution. One fellow with cement burns found cider to be an effective treatment. If you haven't got that, use a bar soap such as Dove which has a low pH. Gojo Orange is an excellent liquid hand cleaner, also with a relatively low pH.

Because the high calcium level in Portland cement mortar leaches moisture from the skin, it is tempting to apply petroleum jelly, a lotion like lanolin, or other skin softening cream, but these products can trap the caustic solution on the skin's surface, where it can cause more damage. Instead, try using aloe vera gel. It helps to soothe the skin and acts as an antiseptic.

12.6: *These are my homemade half-inch mesh and quarter-inch mesh screens.*

Use strong, cloth-lined rubber gloves. Most of our students get used to working with the gloves in a day or two. You have to do it. I knew a stone mason whose hands were always covered with boils because he hated working with rubber gloves. He died too young of skin cancer. And when your gloves get old and don't respond well to a thorough washing, get new ones. It is a good idea to remove jewelry and watches in or near the glove, otherwise wet mortar can get trapped against the skin.

If you get cement in your eyes, do a 15-minute rinse with clean tap water and seek medical attention.

Similarly, cement and lime dust are not good to inhale into the lungs. Use a good respiratory mask for mixing mortar.

Wheelbarrow

For 40 years Jaki and I have been mixing our "mud" in a wheelbarrow. Why not a mortar mixer? Well, a wheelbarrow is quieter and does not upset the karma of the site. Quality control is easier with a wheelbarrow. A wheelbarrow is less expensive. A mortar mixer — as Murphy's Law tells us — breaks down. Finally, to get the mortar to the wall, you need a ... wheelbarrow! Now, it is true that hand-mixing may cause you to rediscover some muscles you've forgotten about, but, in the long run, you will be fitter for the effort. I'm 70, and I still prefer a wheelbarrow to a mixer.

At our workshops, we use two wheelbarrows because we need to keep the mortar coming for 10 or 12 students. Mom and Pop may be fine with a single barrow, but an organized work party will benefit from a second one.

Screens

Sometimes you can buy screens ready-made at the building supply, but I make my own from short lengths of 2×2 or 2×4 stock screwed together. I buy small pieces of the required wire mesh at the hardware store and fasten them to the frame with roofing nails. I use a quarter-inch screen for the sawdust that will be used in the mortar as a cement retarder, and half-inch mesh to screen the sawdust used as insulation.

Two Shovels ...

... of the same size. I like long-handled garden spades for measuring out the constituent ingredients. You need two because one is for the wet sawdust and the other is for the dry goods.

Hoe

A lightweight inexpensive garden hoe works very well for mixing cordwood mortar. You don't need one of those big mason's hoes with the two holes in the blade that professionals use to mix brick mortar or plaster in quantity.

5-Gallon Buckets

Buckets are handy for carrying and storing water, as well as cleaning hoes and shovels. I keep my mixing hoe standing in a bucket of water between mixes.

Hose for Water

If you are fortunate enough to have tap water on site.

Scrubbing Brush

We use a stiff-bristled plastic scrubbing brush to clean the wheelbarrow and any other tools that have contacted mortar, lime, or cement.

Tools for Laying up the Cordwood Masonry

Mortar Pans

We use small, round (12-inch diameter) rubber pans, such as the kinds used for oil changes or as feed pans. Try the farm supply store. We found cheap ones at the dollar store, but they barely lasted a season. You get what you pay for. Old metal pizza pans will work, too, but nothing made of wood, which dries the mortar prematurely.

Rubber Gloves, Again

Their protective quality cannot be over-emphasized. We also use them — not a trowel — for placing mortar on the wall. Jaki and I built our first cordwood home with a trowel, but the late great "Cordwood Jack" Henstridge showed us how much faster and easier it is to place the mortar with gloves.

Pointing Knives

You'll need a few pointing knives. The tools made for brick and block raking are not suitable. They are designed for straight ⅜-inch mortar joints. Jaki and I continue to get inexpensive pointing knives from thrift stores and garage sales. We also look for non-serrated butter knives, like Grandma used to have. We like the ones that are almost an inch wide, but it is good to have a variety, and we even keep one or two with narrower blades for use where log-ends were laid too close together. Bend the last inch or two of the knives to about a 15- or 20-degree angle; that way, you can get the business end in close to the work without your knuckles getting in the way. A good way to bend the tip is to put the tip under a post of your timber frame and gently raise the handle, a little at a time until you have the required angle.

M-I-M-stick. See Sidebar.

Flat and Pointed Trowels

Trowels are useful at the top of a wall, where the cordwood masonry is getting close to the girts of your timber frame; it is hard to put the mortar into that last gap with gloves. Load the back of the trowel with mortar and push it into the gap with a pointing knife. Easy!

Small Buckets and Tin Cans

These are used to pour the sawdust insulation between the inner and outer mortar joints.

Hammer and Chisel, Axe

Useful for splitting off a part of the log-end to make it fit in a space, most often near the top of a panel or when working up to a window buck (frame) already fixed in place.

Four-Foot Level, Mason's Line

Used to keep the wall going up straight, as seen in Chapter 8.

Tape Measure

Used for getting window bucks in the right place, or positioning design features or specialty log-ends.

M-I-M Sticks

You'll want three or four of these on hand. M-I-M stands for *Mortar-Insulation-Mortar.* Make your own M-I-M sticks from 1-inch-thick boards, two or three inches wide. They are used to assist the builder in placing the right width and thickness of mortar for the thickness of wall you are going to build. The top of the stick shows the widths for mortar(s) and insulation. The one-inch dimension of the stick can be used to check the thickness of the joint.

Figure 12.7 shows, top to bottom, an 8-inch M-I-M, a 12-inch M-I-M, foot rule for scale, a 16-inch M-I-M for load-bearing walls, and a 16-inch M-I-M for cordwood within a post-and-beam frame. All are made from smooth one-by boards. All of the M-I-M sticks shown show the quarter-inch "reveal" at each end, indicating that the wood (log-end) is "proud" of the mortar. Below are M-I-M dimensions for various thicknesses of wall, where LB indicates *load-bearing*, and NLB indicates *non-load-bearing.* The M-I-M totals are a half-inch less than the wall thickness. The half-inch is divided between the quarter-inch interior reveal and the quarter-inch exterior reveal, and I draw these reveals right on the stick as a reminder.

12.7: *M-I-M* (Mortar-Insulation-Mortar) *sticks.*

All 8″ walls:	M: 2.5″ I: 2.5″ M: 2.5″
All 12″ walls:	M: 3.75″ I: 4″ M: 3.75″
16″ LB:	M: 4.75″ I: 6″ M: 4.75″
16″ NLB:	M: 3.75″ I: 8″ M: 3.75″
24″ LB:	M: 5.75″ I: 12″ M: 5.75″
24″ NLB:	M: 4.25″ I: 15″ M: 4.25″

Chapter 13

Conclusion

Cordwood masonry is a very old, long-established building method that has been growing in popularity since the 1970s. It has a distinctive character, combining the warmth of wood with the pleasing textural relief of stone masonry. Since 1994, new and improved techniques have been demonstrated at five Continental Cordwood Conferences, and it is hoped that the next gathering will be a *World* Cordwood Conference, perhaps in Europe, where cordwood construction has been enjoying a resurgence over the past several years.

The work is easily learned, but that's not to say that it can't be done poorly or screwed up entirely. It happens. But reading this book is a big step in the right direction. Now you need to actualize the lessons by building a small sample wall or test panel to find out if cordwood is really for you. Do not be discouraged by your speed of building, or lack thereof. Speed — and build quality — comes with practice. A cordwood video can help show efficiency and skill techniques, but it is even better to attend a cordwood masonry workshop with experienced instructors. After just three days of personalized instruction, your initial speed and build quality will be unrecognizable.

Done correctly, a cordwood building will remain beautiful for a very long time. Exterior walls will weather nicely, and the interior should maintain its original appearance decades after construction. Problems of wood and mortar shrinkage can be greatly minimized by paying attention to the best practices outlined in this book. Jaki and I have never had to do wholesale renovation to a cordwood wall; just a few shrinkage gaps to take care of in large special-feature log-ends. Chapter 10 tells what to do about wood or mortar shrinkage, should either come up.

Cordwood homes are bought and sold. How their resale value compares to similarly sized houses is difficult to say, although I'm sure it depends on how well they were built. On the plus side, because cordwood homes rarely involve a mortgage, resale means cash money in the pocket.

Now that you've come this far, I suggest you revisit Chapter 3, Appropriate Use, before starting a project. It should mean it bit more to you now.

Happy stacking!

Glossary

AIR INFILTRATION: The transfer of air though the fabric of a building. Log-ends that shrink a lot are sources of air infiltration and thus promote heat loss by convection.

Bed: In masonry, the mortar upon which a brick, block, stone, or log-end is laid.

Bonding agent: A liquid product made by various manufacturers designed to facilitate the bonding of mortar or plaster to other clean surfaces. See Resources.

Bottle-end, bottle-log: Two bottles or jars joined together to make a glass masonry unit for the admission of light. See Chapter 9.

Built-up corners: In cordwood masonry, a corner system by which corners are constructed of regular wooden blocks, called "quoins," laid up in an alternating crisscross fashion. Also known as "stackwall corners." See also "Lomax corners."

Cement retarder: One of a number of commercially available products used as additives to concrete or mortar for the purpose of slowing the set of the material.

Cement: The hardening and strengthening agent in mortar and concrete. See also "Portland cement" and "masonry cement."

Checking: The natural splitting of a log-end (or any piece of wood), resulting from rapid drying. A presplit log-end often has hairline cracks — without a primary check that goes all the way through from end to end. A single large check is a common condition with cylindrical log-ends.

Cob: A mixture of sand, clay, straw, and water used to build walls. Can also be combined with log-ends to build a cordwood wall. See next entry.

Cobwood: A new term coined by cordwood and cob builders, referring to a cordwood masonry wall tied together with cob instead of mortar.

Concrete: A mixture of sand, stone aggregate, Portland cement, and water. When concrete sets, it makes a strong wall, slab, deck, or foundation material. Not to be confused with "mortar."

Cord: A unit of measure for stacking and purchasing firewood or pulpwood. While, technically, a cord of wood should refer to a "true," "full," or "real" cord of 128 cubic feet, the term now commonly refers to any stacked pile of wood with a sectional area of 32 square feet, normally four feet high and eight feet long. If the stack is also four feet wide, it will be a true cord of 128 cubic feet. See also "face cord."

Cordwood masonry: A wall-building system in which short logs, often called "log-ends," are laid up transversely in the wall within a special mortar matrix, much as a cord of firewood is stacked. Also, "stovewood masonry," "stackwall," "firewood wall," and the like.

Double wall technique: A thermally efficient wall system, made from two separate cordwood masonry walls separated by a fully insulated cavity. Although at least one double wall barn from the 1930s has been identified in Michigan's Upper Peninsula, double wall for housing was developed by Cliff Shockey in 1977. See Chapter 6.

Drawknife: A sharp, single-edged metal blade with a handle at each end of the cutting edge. Used mainly for shaping wood, a drawknife can also make a good barking tool when all else fails. See Chapter 12.

Face cord: A stack (also "rank," "rick," or "run") of wood four feet high, eight feet long, and a certain agreed-upon thickness: 12 or 16 inches, for example. The face cord is a convenient measure to use in determining material requirements for a cordwood project. It is important that the buyer knows exactly what the seller means by the term "cord."

Firewood walls: An archaic term for cordwood masonry walls.

Floating ring beam: A ring of concrete footings floating on a pad of percolating material. See next entry.

Floating slab: A foundation method, whereby a concrete slab is "floated" on a pad built up from runs of good percolating material such as coarse sand, gravel, or crushed stone. A favorite of Frank Lloyd Wright, the floating slab is an economic choice for a cordwood foundation in areas of deep frost. Not recommended on expansive clay soils.

Footer, footing, foundation: A base for a wall or building.

Girder: A major horizontal beam that supports floor joists or roof rafters.

Girt: A beam that joins the tops of sidewall posts around the perimeter of a timber frame structure.

Lime: A white alkaline powder added to mortar to improve its plasticity. "Mason's lime" (also called "builder's lime," "hydrated lime," or "Type S lime"), is made by converting limestone by heat. "Agricultural lime," which is non-hydrated, is used as a soil conditioner in agriculture and is not suitable as a mortar additive.

Lintels: Wooden timbers that carry wall load over doors or windows.

Log-ends: The individual short logs, butts, blocks, ends, or pieces of wood used as masonry units in a cordwood wall. Log-ends are most commonly used transversely in the wall, where, with their end-grain exposed, they "breathe" very well, greatly reducing the danger of wood deterioration through rot.

Lomax corners: Built-up corners made from pre-built corner units, instead of individual quoins. Named for Gary Lomax. See Chapter 6 for details.

Masonry cement: A cement and lime mixture that has become popular with modern masons. There are several types, with varying characteristics.

Mortar mix: Common term for a dry, bagged, premixed mortar product, usually about three parts sand to one part masonry cement. Just add water for a good brick or block mortar. Not to be confused with bags of masonry cement, which contain no sand.

Mortar: A mixture of sand, cement, and water used for laying up masonry units such as bricks, blocks, stones, or log-ends. Sometimes other ingredients are added for certain purposes. Colloquially know as "mud."

Mud: Slang for "mortar."

Panel: A section of masonry enclosed within a timber frame.

Papercrete: A material made from paper, cement, and water, used for building. The density and strength of papercrete varies widely with the recipe and whether or not sand or other admixtures are included. See also "paper-enhanced mortar (PEM)."

Paper-enhanced mortar (PEM): A mortar with a high recycled-paper content. See Chapter 5.

Peeling spud: A chisel-like tool made for removing bark. Many cordwood builders have made successful spuds by mounting a wooden handle to the leaf spring from an old truck. A heavy pointed mason's trowel makes a pretty good peeling spud, too.

Plate beam: In a post-and-beam frame, the topmost horizontal member; the top of a cordwood masonry panel. See also "girt" and "plate."

Plate: Wooden planking (typically two inches thick) used to distribute joist or rafter load onto the cordwood wall. The plate can also tie corners together and provide a surface upon which to fasten floor joists or rafters. A double-thick plate with staggered joints is best.

Pointing knife: A tool used for pointing. Can be made by bending the last inch or two of a smooth kitchen butter knife to an angle raised about 15 degrees from the plane of the knife blade.

Pointing: The process of smoothening the mortar between masonry units. Also called "tuck-pointing" or "grouting." With brick or block work, the term "raking" is also used.

Portland cement: A strong, unmixed cement used in concrete, made by burning a mixture of limestone and clay or other materials. Type I is the basic type, with standardized strength characteristics. Type II is almost the same, but is air-entrained.

Proud: The opposite of recessed. Protrusive: the log-ends sit proud of the mortar background.

Quoins: In cordwood masonry, the individual blocks of wood used in the construction of built-up corners, usually made from regular dimensional material, such as six-by-six-inch timbers. In stone masonry, squared stones used in corner construction.

Random rubble pattern: The use of a variety of sizes and shapes of log-ends, distributed randomly in the wall.

Rank, rick, run: See "face cord."

Retarder: See "cement retarder."

Ridgepole, ridge beam: The major carrying beam or girder of a roof system, supported by posts.

R-value: A measure of insulation value in building materials. The higher the R-value number, the greater the insulation. Materials are often measured in terms of R-value per inch. Extruded polystyrene, for example, is about R-5/inch.

Sill plate: A wooden plate, often pressure treated, that caps the top of concrete footings, a poured concrete wall, or a block wall. Also "toe-plate" or "sill."

Sills: Heavy horizontal wooden timbers sometimes installed beneath window framing. See also "sill plate."

Stackwall corners: Same as "built-up corners."

Stackwall: Cordwood masonry, particularly in Canada.

Stovewood masonry: Same as "cordwood masonry." The term is most commonly encountered in historical articles and is seldom, if ever, used in reference to cordwood structures built since 1960.

Thermal mass: The capacity of a material to store heat. Generally, a material's thermal mass characteristics are inversely proportional to its insulation characteristics.

Toe-plate: See "sill plate."

Resources

Products

Cement Retarders

I HAVE HAD GOOD RESULTS with the first two listed below. A five-gallon drum of Sika Plastiment lasted me for years. Your local building supply may have others not listed here.

Sika Corporation. "Plastiment is a water-reducing and retarding admixture." Sika has offices all over the world. Go to sika.com and search for your country. In the United States: Sika Corporation, 201 Polito Avenue, Lyndhurst, NJ 07071. Tel: 800-933-7452. Website: usa.sika.com.

W.R. Grace and Company. 62 Whittemore Avenue, Cambridge, MA 02140. Tel: 617-876-1400 (24 hours) or 800-354-5414 (8 am–5 pm Eastern). Website: grace.com. Makes W.R. Grace Daratard 17: "An admixture for use where delay in setting time is required to ensure sufficient placement, vibration or compaction time. Comes in 55-gallon drums." W.R. Grace makes lots of other cement retarders. Search their site. (Tip: Look for Grace retarders at your local concrete batch plant.)

Increte Systems. 1611 Gunn Hwy., Odessa, FL 33556. Tel: 813-886-8811 or 800-752-4626 Website: increte.com. "Increte Systems cement retarder is an easy-to-use, water-based additive to prolong the setting time for cement products. Use any time longer set times are desired."

Bonding Agents

DAP Products, Inc. 2400 Boston Street, Suite 200, Baltimore, MD 21224. Tel: 410-675-2100. Website: dap.com. "DAP Bonding Liquid and Floor Leveler Additive is a versatile product that can be used as an additive to increase adhesion and flexibility in plaster, mortar, concrete, floor leveler, stucco and weatherproof cement paint."

Thoro Consumer Products. BASF Corporation Building Systems, 889 Valley Park Drive, Shakopee, MN 55379. Tel: 216-839-7171 or 866-518-7171. Website: thoroproducts.com. Makes Acryl-60 Bonding Agent: "Acrylic, polymer emulsion additive for cement-based powders designed to improve adhesion, tensile, compressive, and flexural strengths. Non-yellowing, water-based ideal for both interior and exterior use."

Spray Foam Insulation

Here are three of the leading players in spray foam insulation in the United States and Canada. Contact them to find a contractor near you who can apply their product. You can find many others with an online search. Also, check directly with your local insulation installers.

BioBased Technologies. 3333 Pinnacle Hills Parkway, Suite 400, Rogers, AR 72758. Tel: 877-476-5965. Website: biobsed.net.

Icynene. 6747 Campobello Road, Mississauga, Ontario, L5N 2L7, Canada. Tel: 800-758-7325. Website: icynene.com.

Demilec, Inc. 3315 E. Division Street, Arlington, TX 76011. Tel: 888-224-1533. Website: demilec.com. In Canada: Demilec, Inc., 870 Cure Boivin, Boisbriand, Quebec, J7G 2A7, Canada. Tel: 866-345-3916. Website: demilec.ca.

Sources

(*Author's note:* I have confined this list to materials current at press time.)

Cordwood Books

Flatau, Richard. *Cordwood Construction: Best Practices.* Cordwood Construction Resources, 2012, revised and updated 2015. True to its title, the author details "best practices" methods about cordwood masonry and its relationship to foundations, electrical considerations, energy codes, and more. It's a well-illustrated and meticulously documented work. 196 large 8.5" × 11" pages, including 259 color pictures and diagrams.

Roy, Rob. *Cordwood Building: A Comprehensive Guide to the State of the Art.* New Society Publishers, 2016. This fully revised edition of the most widely read book on cordwood construction presents the latest innovations. It differs from the book in your hand in that it has nine comprehensive case studies from around the world and a detailed history of cordwood masonry. 264 pages, plus an 8-page color section that is completely different from the one herein.

— — —, *The Sauna.* Chelsea Green, 2004. A cordwood masonry sauna is a great starter project and delivers a genuine Finnish sauna experience. This book is about saunas, but three of its nine chapters are really about cordwood. Chapter 4 is about building a post-and-beam log-end sauna and Chapter 5 is about building a round cordwood sauna. Other chapters deal with sauna lore, siting and design, stoves, and how to take a sauna. Fully illustrated, 236 pages. Although out of print, this book is still available from Earthwood at cordwoodmasonry.com.

— — —, *Stoneview: How to Build an Eco-Friendly Little Guesthouse.* New Society Publishers, 2008. Stoneview is an octagonal cordwood masonry timber-framed guesthouse with a living roof, located at Earthwood. Over 130 clear line drawings and step-by-step images detail all the information needed to build Stoneview from start to finish and a color section shows the design features of this charming "green" cabin. All design considerations are covered, as well as a thorough discussion of octagon geometry. Chapters are devoted to site prep, forming and pouring the slab, timber framing, the lightweight living roof, and the cordwood walls. 244 pages, four in full color.

— — —, *Timber Framing for the Rest of Us.* New Society Publishers, 2004. Many natural building methods rely on a timber frame first, which is then in-filled with straw, cob, cordwood, or more conventional wall materials. But traditional timber framing employs the use of finely crafted joints and wooden pegs, requiring a high degree of craftsmanship and training, as well as much time and expense. This book describes the timber framing methods used by most contractors, farmers, and owner-builders: methods that use modern metal fasteners, special screws, and common-sense building principles to accomplish the same goal in much less time. This is the first book to describe in depth these more common fastening methods. 176 pages, well illustrated.

Stankevitz, Alan, Richard Flatau, Rob Roy, and Dr. Kris Dick. *Cordwood and the Code: A Building Permit Guide.* Described in Chapter 11 herein.

Cordwood Videos

Flatau, Richard. *Cordwood Construction Best Practices DVD.* There are 30 menu items: Wood selection, R-values, mortar mixes, building a wall, pointing, best practices, electrical, plumbing, foundations, framing, special effects, and more. Available at cordwoodconstruction.org

Roy, Rob and Jaki Roy. *The Complete Cordwood DVD,* an Earthwood Building School/ Chevalier-Thurling Productions video. This DVD combines *Cordwood Homes,* a tour of eight cordwood homes around North America with our *Cordwood Masonry Techniques,* filmed at our Earthwood workshops. An extensive captioned slideshow rounds out the DVD, showing examples of round, post-and-beam, and stackwall buildings, as well as construction details and special features. 3.25 hours. Available at cordwoodmasonry.com

Cordwood Masonry Websites

cordwoodmasonry.com

Our Earthwood Building School website. Since 1979, Earthwood has been conducting cordwood workshops in northern New York and around the world. The site serves as a clearinghouse for all things cordwood, including the books in this bibliography.

cordwoodconstruction.com

Richard Flatau and his wife Becky conduct cordwood masonry workshops in Wisconsin and other states. Richard is the author of *Cordwood Construction: Best Practices.* His website has a wealth of interesting cordwood pictures and articles.

There are many more websites and blogs, too numerous to list, that tell of individual experiences with cordwood masonry. Simply plug "cordwood masonry" or "cordwood building" into a search engine and be prepared for a long and interesting afternoon. One to try, though, is daycreek.com by Alan Stankevitz.

Related Websites

greenhomebuilding.com

Kelly Hart's website covers all sorts of natural and vernacular building, including a section on cordwood masonry. In the "Ask Our Experts" section, Rob Roy fields the cordwood questions, and hundreds of Q&As are listed, with categories like: types of wood to use; where to find cordwood; debarking and curing the wood; foundations; appropriate mortar and methods; structural considerations; code and permit issues.

newsociety.com

This is the site for New Society Publishers. If you have enjoyed this book, you'll probably be interested in the other titles in their Sustainable Building Essentials series.

Index

Page numbers in *italics* indicate tables.

About the Author

Rob Roy was born in Webster, Massachusetts, in 1947. He lived seven years in the Scottish Highlands, where, in 1972, he met Jaki, now his wife of 45 years. The couple has lived in West Chazy, New York, since 1975.

Rob is the Director of Earthwood Building School, which he co-founded — with Jaki — in 1981. Earthwood teaches owner-builders how to build with cordwood masonry and "timber framing for the rest of us" techniques, making use of commonly available fasteners. Rob and Jaki also teach earth-sheltered housing — including living roofs — and megalithic stone raising using the same materials that were available to people 5,000 years ago. Rob has written 15 books in these and related fields, including *The Sauna* and *Mortgage Free!* His New Society titles include *Cordwood Building: A Comprehensive Guide to the State of the Art* (2016), *Timber Framing for the Rest of Us* (2004), *Earth-sheltered Houses* (2006), and *Stoneview: How to Build an Eco-friendly Little Guesthouse* (2008).

Rob and Jaki have built four cordwood homes and numerous outbuildings, and have conducted workshops all over North America and around the world. Their love of travel resulted in their first co-authored book, *The Coincidental Traveler: Adventure Travel for Budget-minded Grown-ups* (Earthwood Publishing, 2014), in which they share strategies and philosophies they use to create memorable existential travel experiences.

A Note About the Publisher

NEW SOCIETY PUBLISHERS is an activist, solutions-oriented publisher focused on publishing books for a world of change. Our books offer tips, tools, and insights from leading experts in sustainable building, homesteading, climate change, environment, conscientious commerce, renewable energy, and more — positive solutions for troubled times.

DON'T EAT THIS BOOK *(but you could)*

We're proud to hold to the highest environmental and social standards of any publisher in North America. This is why some of our books might cost a little more. We think it's worth it!

- We print all our books in North America, never overseas
- All our books are printed on **100% post-consumer recycled paper**, processed chlorine free, with low-VOC vegetable-based inks (since 2002)
- Our corporate structure is an innovative employee shareholder agreement, so we're one-third employee-owned (since 2015)
- We're carbon-neutral (since 2006)
- We're certified as a B Corporation (since 2016)

At New Society Publishers, we care deeply about *what* we publish — but also about how we do business.

New Society Publishers

ENVIRONMENTAL BENEFITS STATEMENT

For every 5,000 books printed, New Society saves the following resources:[1]

31	Trees
2,784	Pounds of Solid Waste
3,064	Gallons of Water
3,996	Kilowatt Hours of Electricity
5,062	Pounds of Greenhouse Gases
22	Pounds of HAPs, VOCs, and AOX Combined
8	Cubic Yards of Landfill Space

[1]Environmental benefits are calculated based on research done by the Environmental Defense Fund and other members of the Paper Task Force who study the environmental impacts of the paper industry.

Certified B Corporation

FSC www.fsc.org MIX Paper from responsible sources FSC® C016245

new society PUBLISHERS www.newsociety.com